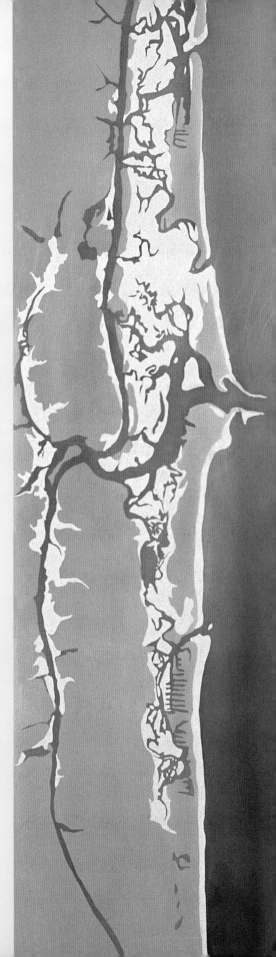

A CELEBRATION OF THE WORLD'S BARRIER ISLANDS

COLUMBIA UNIVERSITY PRESS
NEW YORK

A CELEBRATION OF THE WORLD'S

Barrier Islands

Text by

ORRIN H. PILKEY

Original Batiks by

MARY EDNA FRASER

♕

COLUMBIA UNIVERSITY PRESS

Publishers Since 1893

New York   Chichester, West Sussex

Library of Congress Cataloging-in-Publication Data

Pilkey, Orrin H., 1934–

    A celebration of the world's Barrier islands / by Orrin H.

Pilkey ; original batiks by Mary Edna Fraser.

      p.   cm.

    Includes bibliographical references (p.).

    ISBN 0–231–11970–4 (cloth : alk. paper)

    1. Barrier islands.   I. Fraser, Mary Edna.   II. Title.

GB471 .P63 2003

551.42'3—dc21

2002071593

∞

Columbia University Press books are printed on permanent and durable
acid-free paper.

Printed in China

c 10 9 8 7 6 5 4 3 2 1

To Sharlene and John
Our Children
and Our Grandchildren

CONTENTS

ILLUSTRATIONS

Figures

THE idea for this book was born when Mary Edna Fraser and I met on a research trip to Cape Lookout National Seashore in 1993. Her exhibition at the Smithsonian National Air and Space Museum in 1994–1995, "Aerial Inspirations: Silk Batiks by Mary Edna Fraser," needed wall text explaining the islands shown in her art, and she wanted my input. We soon realized that we shared a vision of preserving the barrier islands. Above the roar of the ship's engine as we crossed the inlet, we shook hands and agreed to collaborate. We would communicate our vision by combining her aerial photographs and silk batiks with my knowledge of the science of barrier islands.

Mary Edna and I hoped that if people understood what barrier islands are and how they operate, they might behave differently. Translated into political terms, what we hoped for was that people who were educated on the subject might support stronger laws and regulations that would allow the islands to survive for future generations to enjoy. It's not too late to accomplish this goal. Surprisingly enough, most of the world's barrier islands are uninhabited. Some, like those in Siberia and on the North Slope of Alaska, are uninhabited because of their harsh climate. Others, like those in Colombia, are uninhabited because they are too dynamic. Some barrier islands, however, in such diverse locales as Brazil, Nigeria, Portugal, Holland, Mozambique, Egypt, and Ecuador, are viewed as potential hot spots for attracting tourist dollars.

Looking beyond the politics, the rush to develop, and the huge risks to life and property on the islands, one realizes that they are a fascinating phenomenon. Islands make up perhaps 12 percent of the world's open ocean shorelines, and "new" island chains are still being recognized, especially along tropical coasts. A great deal of fine-tuning in the science of barrier islands has yet to be accomplished, and scientists must iron out many areas of disagreement, especially concerning the rates of island evolution, the ages of past island events, and the ways in which the islands will respond to the coming rise in sea level. We do, however, understand the basic mechanics of island evolution, and the time is ripe for this book.

Mary Edna and I believe that barrier islands are akin to coral reefs. Both are endangered by the activities of humans, and both are essentially irreplaceable once they are lost. Whereas coral reefs can be killed almost overnight, barrier islands take decades, even a century, to die. It is much easier to be concerned with a short-term problem than with one that may occur when the next generation is in charge, but if the islands are to be preserved or sensibly developed, we must take the long view. The barrier islands of the world are every bit as important as coral reefs.

FIRST and foremost I want to thank Virginia Henderson. She made the first study of the global distribution of barrier islands within our program at Duke University (and one of the first in the world). When she began her master's work, we naively thought that all we would have to do would be to open up a few world atlases and, bingo, we could complete our survey. Now, well over a decade later, we still haven't nailed down some barrier islands. We have discovered "new" chains of islands (although we are sure we haven't found them all) and have learned all kinds of things about the uncertainties and omissions of maps, charts, and satellite images. Virginia's doggedness and never-ending curiosity led her to make huge strides toward a global understanding of barrier islands. I am indeed sad to note that Virginia died too young in 2001, a victim of breast cancer.

This book would not exist without the kick-start that Virginia gave it. She passed the baton to Matt Stutz, who is completing his dissertation on a global view of barrier islands. Matt's studies have provided much information and a great deal of understanding of the barrier island big picture all around the world. I also want to give a special acknowledgment to Miles Hayes, whom I consider to be the scientific "king" of barrier islands. He contributed much intuition and many slides. As always, I am grateful for the consultation and editing help of my longtime friend and coastal study associate Bill Neal. Bruce Molnia helped us navigate the political shoals of satellite imagery.

Holly Hodder, our first editor at Columbia University Press, set the tone for the book's organization and quality. Early chapter drafts were reviewed by Monte Basgall, Bill Neal, and Kathy Dixon. I even called in my family to help with this task, including my son, Keith; my daughters, Linda and Diane; and my wife, Sharlene. Thanks to Amber Taylor for drafting and to Jan McInroy for copyediting.

This book takes a global view, and people all over the globe helped us to understand their islands. Without their assistance, the project would have been as dead as a barrier island in Taiwan. Many individuals provided us with advice, intuition, photos, and studies, often unpublished. In our research on U.S. barrier islands we received state-by-state assistance from Bob Morton in Texas; Greg Stone in Louisiana; Richard Davis and Charles Finkl in Florida; Jim Henry and Buddy Sullivan in Georgia; Gered Lennon and Paul Gayes in South Carolina; Stan Riggs, Bill Cleary, Rob Thieler, and Duncan Heron in North Carolina; Henry Dixon in Virginia; Chris Kraft in Delaware; and Bill Schwab and Sue Halsey in New York and New Jersey. For assistance with Arctic barrier islands in Canada, the United States, and Siberia, we thank Erk Reimnitz, Jess Walker, Owen Mason, Andy Short, Bob Taylor, Art Trembanis, Matt Stutz, and Steve Solomon. On the international scene, we are deeply grateful to the following people: Jaime Martinez and Juan Gonzalez in Colombia; Gene Shinn in Abu Dhabi; Ramon Gonzalez, Joao Dias, and Jose Monteiro in Portugal; Al Hine, Dag Nummedal, and Richard Anderson in Iceland; Andrew Cooper in Mozambique; Sam Smith, Andy Short, and Bruce Thom in Australia; Dave Bush, Bill Neal, and Rob Young in Mexico; Dan Stanley in Egypt; Tsung Yi Lin in Taiwan; and Albert Prakken, Rudi Stanhoff, and Saskia Jelgersma in Holland.

Mary Edna would like to thank John Sperry, Sarah and Rebecca Fraser, Mary and Rebecca Burkhead, and Laura Faver for their input and support; Claude Burkhead Jr. and Claude Burkhead Sr. for piloting, keeping the Ercoupe airworthy, and providing the fun of this aerial adventure; and Rick Rhodes and Terry Richardson for photographing the batiks. She is grateful to Mary Quattlebaum, Blair Lambert, Patti Holsclaw, and Brenda Drain for assistance in the studio and with exhibitions; Leslie Sautter, Diana L. Turnbow, and family members for editing; and Nancy Sears for public relations. We are both grateful to the Duke University Museum of Art, the National Science Foundation, the National Academy of Sciences, the North Carolina Maritime Museum, the United States Geological Survey, East Carolina

University Gray Gallery, the Gibbes Museum of Art and the Dayton Art Institute for exhibitions and for playing a role in refining our partnership.

Special thanks go to Susan Lawson-Bell for curatorial advice and to Greg Bryant for production of the innovative CD-ROM of our work. Priscilla Strain and Carolyn Russo at the Smithsonian National Air and Space Museum fostered the project from the start. Thanks to my dear friends Marjory Wentworth, the Anonymity Dance Troupe, Dana Downs, the late Frances Morgen, Rich Robinson, Clay Taliaferro, Mitchell Davis, and Dorsey Worthy for their creative collaboration. Thanks as well to Jim Nicholson, Tsung-Yi Lin, Matt Stutz, Daniel Stanley, Erk Reimnitz, Dag Nummedal, and all of the scientists who provided reference for the illustrations. Thanks also to Diane M. Boltz, Hank Buchard, Michael Kilian, and Patty Kim, as well as every journalist who thoughtfully wrote about our passion for the islands and bolstered this project. Anne Collins Goodyear, Jeff Kopish, and Marie Pelzer have been artistic consultants in this venture. I appreciate as well the help of the South Carolina Arts Commission; Beryl Dakers and Elaine Cooper at South Carolina Educational Television; and Gil Shuler Graphic Design for decades of assistance.

Thank you to all of my patrons through the years, especially Norton and Mindy Seltzer, the American Embassy of Thailand; Rosamond Talbe at Computer Sciences Corporation (CSC), Columbia, South Carolina; the Gibbes Museum of Art, Charleston; the Medical University of South Carolina; and Sheila David at the H. John Heinz III Center for Science, Economics, and the Environment in Washington, D.C. I am also grateful to Rindy and Mark Abdelnour, Ingrid Abendroth, Joan Avioli, Courtney and Phillip Flexon, Burke Graham, Eunice and Herman Grossman, Claire Gummere, Marianne and Martin Harwit, Margaret Pinckney Hay, Meg Hoyle, Lisa and Robert Irvin, Pamela and Stan Kaplan, Barbara and Matt Kimbrell, Linda and Phil Lader, Margaret and Bobby Minis, Angela and Roger Morrow, Kimberly and Jeffrey Newhouse, John Sanders, Mary Trent Semans, Elliott Steadman, Gail Stuart, Kaye and Johnny Wallace, Ruth and Larry Willis, Cheryl VanLandingham, and Tina and Larry Vertal.

A CELEBRATION OF THE WORLD'S BARRIER ISLANDS

HURRICANE Dennis (1999), which will be described in chapter 1, was a single, brief, violent event in the life of the Outer Banks barrier island chain of North Carolina. There are thousands of such events to come for mobile islands like these all over the world. More storms are a certainty. A rising sea level is a certainty, too. And so are increased pressure to develop the land, more dams on the rivers that supply sand, deeper navigation channels between the islands, and more seawalls to lock up sand. One could say that the barrier islands of the world, arguably the most dynamic geologic features on the surface of the globe, have a most exciting future.

As Mary Edna Fraser and I plunge into this world of barrier islands it will soon become obvious that there are as many ways in which barrier islands evolve as there are barrier islands. No two islands have the same orientation, the same waves, winds, and tides, the same size and frequency of storms, the same sea level rise, the same vegetation, and the same sand supply. If we understand how the barrier islands of the Outer Banks evolve, that doesn't mean we understand the mechanics of tropical Colombian islands, ice-encased Siberian islands, the desert islands of Abu Dhabi, or the delta islands of the Niger River in Nigeria.

The islands in Colombia are affected by the high sea levels and increased storminess during El Niño weather events and have a huge sand supply rushing down the slopes of the nearby Andes Mountains. Waves that shape Siberian Arctic islands can work only during the two ice-free summer months. Much of the sand that makes up the islands of Abu Dhabi is calcium carbonate precipitated out of seawater. The Niger Delta islands suffer from sand starvation because of sand trapping by upstream dams.

Although all these islands are vastly different in many ways, they have much in common. Each is a pile of unconsolidated sand or sometimes gravel, longer than it is wide. In front of each is an ocean, and behind it is a lagoon. They are called barrier islands because they act as outer barriers protecting the mainland shorelines behind them from the ravages of storm waves and tsunamis. While that is a fine, useful, even admirable function from the standpoint of humans, protection of the mainland is not the reason that barrier islands form. Nature creates islands because they create the most efficient edges of continents, a way of coming close to a line of sand that is neither eroding nor building up.

In the ensuing chapters we will explore *why* barrier islands form and *how* they form. Many theories about their origin have been proposed, and probably all of them are right. At least, they all work somewhere. Some of the earliest ideas concerning the origin of barrier islands sprang from the study of ancient landforms now far from the sea.

What makes barrier islands really different from any other topographic feature on earth is their ability to maintain themselves as a unit as they roll across a flooding coastal plain in response to a rise in sea level. As each island rolls landward, the lagoon is maintained because the mainland shoreline erodes back at a pace similar to that of the island's movement. We will argue that in some cases these islands formed far out on the continental shelves when sea level was lower and migrated into their present location when the glaciers covering the continents melted. The sea level is beginning to rise again now, and the islands are beginning to respond.

We will develop the theme that there are islands and there are islands. We recognize nine types of barrier islands, but most islands fall into three categories. *Coastal plain islands* rim the flat plains bordering the Atlantic and the Gulf of Mexico. The *Arctic islands* present a unique and downright exciting special case of coastal plain islands. Waves, ice, and the melting of underlying permafrost during two months of each year shape them. During the rest of the year, they are solidly encased in ice. *Delta islands* line the margins of the bodies of mud and sand that pour into the sea from rivers. Of these, the Mississippi Delta islands are by far the most studied and the best understood. Other deltas with well-developed barrier island chains include the Niger, the Nile, and the Mekong, and we will examine them all.

In some ways, the rare island types, like the sandur islands of Iceland, are the most fascinating to us. Sandur islands migrate in a seaward direction, the only islands that do. Some barrier islands are widening in place, but no others are actually migrating toward the ocean. Some Australian islands could be called fakes. Though they have all the appearances of barrier islands, the sand that makes up their volume accumulated as dunes on the mainland, not as barrier islands in the ocean. The islands just happen to be a thick pile of dune sand that has been eroded and isolated by the present-day sea level.

People build seawalls to hold barrier islands in place. In the Arctic, *the permafrost*, or (semi)permanently frozen ground beneath the barrier islands until it melts, acts just like a seawall, and in the tropics, *beachrock* is equivalent to the permafrost of the Arctic. Beachrock is beach sand cemented in place by calcium carbonate that precipitated out of seawater, and it behaves like the permafrost that behaves like a seawall; all temporarily retard erosion.

One cannot separate the story of the islands and their natural evolution from that of the people who live on them. In fact, the different modes of coping with dynamic processes on barrier islands are an important and intriguing part of the global barrier island story. Nobody lives in a more rugged environment that the Inupiat Eskimos of the Alaskan Arctic. No longer a mobile people at home in easily moved skin, sod, and whalebone huts, their current barrier island existence in ever more crowded communities has proved almost untenable. In tropical Colombia and Nigeria, poverty-stricken island residents routinely move buildings back from eroding shorelines. In Colombia, for all practical purposes, island living just doesn't work, because of the frequent tsunamis on this earthquake-prone coast, among other reasons. The contrast between these marginal communities and U.S. developed islands couldn't be greater. In this country the islands have developed exactly like mainland communities. When erosion, inlet formation, or storm overwash becomes a problem, the solution is an infusion of money. That approach would be virtually unthinkable for a Colombian or Nigerian island community.

People do not live peacefully with barrier islands. It seems that the richer the country is, the less placid the coexistence. A critical part of the global story of barrier islands is the impact of this unpeaceful coexistence of people and nature at the shore. Finding a solution to the human-caused changes in barrier islands preoccupies many a government entity around the world. Many a politician's career rests on satisfying the varied interests that focus on barrier islands, and more than one political career has foundered in the process.

The very nature of barrier islands, existing as they do in a delicate equilibrium between sea level, sand supply, and waves and currents, makes them extremely vulnerable to human-created changes. Almost without exception the changes are for the worse, at least as far as the island is concerned, for they cause the delicate island evolution equilibrium to go awry.

Navigation channels dredged between barrier islands are probably the most important global cause of sand supply loss to them. This can happen even in remote areas. When a Río Patia channel was relocated to facilitate logging in the Colombian tropics, for example, three small barrier islands that depended on the river for sand completely disappeared. The most spectacular example of human impacts, however, may be on delta islands. Dependent on a continuing supply of sand from the rivers, these islands simply starve when the sand is trapped by upstream dams or by levees and canals on the delta. Islands in the Mississippi Delta and the Nile Delta are the best-known examples. Loss of the islands on the Mississippi will affect thousands of people, and on the Nile Delta it will affect millions. But since the Mississippi problem is in a wealthy country and the Nile problem is in a poor one, the "solutions" will be dramatically different and disproportionate to their global societal importance.

The global story of barrier islands and people is the story of different strokes for different folks, differing national priorities, and differing levels of concern for the preservation of these islands. Taiwan has filled in most of its lagoons and armored the island fronts with massive seawalls, virtually destroying thirty islands, with only three or four remaining before the annihilation is complete. In the Netherlands, where much of the shoreline is also heavily engineered, the Dutch treasure the relatively pristine nature of their five barrier islands, prohibiting beachfront development on them and taking costly steps to preserve them. In almost every developed country with barrier islands, seawalls, breakwaters, and jetties have been placed on shorelines, and hundreds of miles of beaches have been destroyed as a result.

More important in the long run is that the island where the beach is lost and the natural processes halted becomes a pile of lifeless sand. As we can see on Taiwan's east coast and in Seabright, New Jersey, islands stabilized for the pleasure of humans no longer vibrate, evolve, and change. All natural activity has been halted—forever. Remove the seawalls and watch the islands disappear.

This is also the story of the geologists, biologists, and other scientists who study these often remote islands and of their triumphs, adventures, and missteps. Unlike most visitors to barrier islands, the scientists revel in bad weather because barrier islands are more likely to evolve during storms than at any other time. Study of the Arctic coast has proved to be downright dangerous to American, Canadian, and Russian researchers who invade the kingdom of the polar bear. Forays into the tropical rain forests, though much less spectacular, are almost as dangerous because of the swarms of bees and the malaria-bearing mosquitoes. Geologists who study the already developed islands of Florida face the hazard of local police officers, who are doggedly determined to keep out the unwashed masses—i.e., nonresidents. People who wander around on portions of islands that strike the fancy of nobody else inevitably bring out the suspicions of local people. "Trespassing" barrier island scientists have been mistaken for government officials, *bandidos*, environmental extremists, and spies, always assumed to be up to no good!

Our voyage to the twenty-two hundred barrier islands of the world ends with an examination of their state of development. Most are uninhabited, and would-be buyers see in them a wealth of very inexpensive beachfront lots, especially in the tropics and the Arctic. To paraphrase the Florida real estate ads, in Siberia there is a lot with your name on it, a beautiful piece of land with a fantastic view of the sea. We will examine the wisdom of developing these restless ribbons of sand and speculate about the future of islands in a rising sea level. A century from now, it is possible that even the rich societies of the world will release their barrier islands back to the forces of nature. We may abandon Fort Lauderdale in favor of preserving Manhattan.

The barrier islands of the globe seem to be the canaries in the coal mine. They warn us, before other features on the surface of the Earth do, that the sea level is rising, our planet is forever changing, and the good old days when nature seemed cooperative and malleable at the shoreline are gone forever.

My partner and illustrator for this book lives just behind a barrier island near Charleston, South Carolina. She knows all about storms—the most important natural events in the life of barrier islands—because Hurricane Hugo devastated her art studio in September of 1989. In 1999 she spent fourteen hours with her two young daughters, a dog, a cat, and a rabbit in a car that was an integral part of a fifty-mile-long traffic jam of people trying to escape Hurricane Floyd. As it turns out, the behavior of hurricanes is as unpredictable as is the conduct of barrier islands—Floyd never came ashore in South Carolina. These experiences with the hazards of coastal living brought her a step closer to understanding the intertwined dynamics of hurricanes and islands.

In Mary Edna Fraser's own words:

My interest in the fragile ribbons of sand that separate the oceans from the mainland is a direct result of flying with my father and brother as pilots over the barrier islands of the Atlantic Coast. My art is based on twenty-three years of aerial photography, often from the perspective of flying in my grandfather's 1946 415C Ercoupe . . . at about 85 miles per hour, the cockpit open . . . wind in my face. Through the Nikon 35mm camera's eye I have scrutinized most of the eastern and western coastlines of the United States and many aerial landscapes abroad. What I have observed is both breathtakingly beautiful and disturbing. Usually a research excursion over undeveloped aerial landscapes will yield about five hundred photographs, approximately twenty of which will be chosen to translate into my medium of batik on silk. However, some trips of up to eight hours have not yielded a single photo I can use for a design; jetties, seawalls, landfills, and false harbors have altered nature beyond recognition. Orrin Pilkey's early book, *The Beaches Are Moving*, verifies that what I see aesthetically in the air is indeed true from a geological standpoint.

My medium is batik; silk cloth colored by hand using a modern variation of an ancient method of dyeing textiles. I prefer to investigate a region firsthand before beginning a batik: hiking the terrain, exploring the waterways by boat and air, collecting rock and shell samples, and making on-site watercolor studies. Maps and nautical charts provide accurate data with which to plan expansive compositions. I use satellite and space shuttle imagery, as well as photographs furnished by scientists, for distant regions that I cannot photograph from the air myself. Reading the scientific papers and listening to Pilkey's explanations, I begin to focus on an aesthetic organization of the land that best depicts the unique qualities of the islands.

Batik is a process in which removable wax is applied to fabric, creating areas that will resist dye, while unwaxed areas absorb dye. This technique of dyeing cloth predates recorded history. Though its origins are unknown, evidence of its early practice has been found in the Far East, Middle East, Central Asia, Africa, and India. The word *batik* originates from the islands of Java in Indonesia, where the art form flourishes.

To begin a batik, I draw on the silk a general outline in pencil, either freehand or by projecting a slide of the image onto the fabric. I select source materials of photos, maps, scientific illustrations, and charts to include in the image. I then stretch the silk panel horizontally and apply hot, melted wax to resist the first dye application. The fabric must often be waxed on both sides to ensure complete penetration. The wax is half beeswax and half paraffin. A typical large batik, measuring twelve by four feet, may require five pounds of wax.

The tools of the trade include soft-bristle wax brushes and the *tjanting*, a spouted copper bowl with a handle, which is used to create very fine lines with hot liquid wax. I often outline the design using the *tjanting*, then brush in the outlined area with wax. The process requires care and precision, since there is no simple way to erase an unwanted drop of wax or dye. The fiber-reactive proceon dyes react chemically with the silk to become part of the cloth. The dyes come in powder form and must be mixed in exact proportions of water, urea, Calgon, baking soda, and washing soda. Any mistake in the chemistry will cause the dyes to bleed, ruining the batik.

Protective gloves, mask, and a well-ventilated dye room are essential, since in powder form the dyes create tiny airborne particles that are carcinogenic. I test color on paper towels and spend several hours working out a color harmony, often comparing dyes to colors in nature or in the location's culture. I apply the liquid dyes with a brush, and the waxed areas resist subtle color transitions. Each application of wax protects the areas colored by the previous dye bath, leaving exposed colors to react with the new dye bath. It is necessary to know color theory and to think of negative space upside down and backward.

Heat removes the wax and sets the dyes. I sandwich the heavily waxed and dyed batik between two layers of white paper and place it on a pad of newsprint. A common household iron melts the wax, which soaks into the paper. After several ironings when the paper absorbs no more wax, I iron the batik once more to further heat-set the dyes. Dry cleaning removes any wax residue, and washing the batik in my washing machine using a chemical detergent, Synthropol, further removes excess dye and pencil lines. This wash tests for dye permanency, and I iron it while still damp to smooth out the surface. Sewing seams for hanging rods in the top and bottom is the last step before display.

Although my medium is labor-intensive, I enjoy every part of the art from the leap in my heart when I see an amazing shot through the camera lens to the final batik that makes the mind's eye a reality. The slowness of the process encourages a contemplative approach. I think about the research papers I have read, the scientists to whom I have spoken, and the discussions that Pilkey and I have had on the subject. When books on the site's culture are available, I refer to them, studying the art and listening to music indigenous to the people who live on the islands.

My intent is to convey the essence of place. My generation is the first to have this dye chemistry available. We are also the first to have fast film and satellite imagery. The work on this book has been greatly influenced by the leaps made during this era of technological growth.

Everything looks different from an aerial perspective—your viewpoint has the natural curve of the Earth; you can literally see the wave patterns on the beach at low altitudes, the distinct separation of sand, marsh, and forest. As I surveyed Georgia's Sea Islands with my brother, Claude Burkhead, in 1980, it dawned on me that the barrier islands are dynamic individuals. How do they form and evolve? What makes each inlet and island unique? It was more than a decade later before Pilkey began to answer my questions.

Flying the Ercoupe over Kitty Hawk, North Carolina, with my dad in 1993 was the first research trip for *A Celebration of the World's Barrier Islands*. My father and I flew a 172 Cessna to see Orrin in Durham, North Carolina, that February. We pored over charts and targeted twenty areas on the Outer Banks to photograph. The next day, on a dawn-to-dusk flight in the little Ercoupe, I surveyed both overdeveloped and natural islands.

It is difficult to capture what I feel when I'm flying over shifting islands, gaining altitude, descending to photograph a moment of visual poetry. The snaking of tidal creeks, the straight line of the Intracoastal Waterway, the abstract quality of our planet as seen from the air is intriguing. The scientific and visual variety of the Earth's barrier islands has proved to be astounding. The passion we feel for this global phenomenon is contagious.

Aerial photographs tell a story of time. Not geologic time but vast changes that happen in one lifetime. Our ecological awareness should make us all activists. My goal is to use art as a vehicle to make the fragility of barrier islands known as an important environmental concern. The batiks convey perspectives that the human eye, maps, and ordinary cameras cannot reveal. I hope the art will contribute to the appreciation of the dynamic nature of these movable strips of sand and will act as a catalyst for the preservation of barrier islands for future generations.

Mary Edna Fraser and I welcome you to our world of barrier islands. She and I will take you on a long-distance journey from pole to pole, hemisphere to hemisphere, and continent to continent. We will show you barrier islands through the eyes of a geologist and an artist, a combination of practical science and beauty. Perhaps you will come to see these fascinating islands as we do—as beautiful, dynamic, and vanishing.

CHAPTER 1

Dennis Roars Ashore

A Beneficial Catastrophe

I GOT the call from CBS News about two in the afternoon. Hurricane Dennis had just arrived off the North Carolina coast; waves were already washing across the barrier islands of the Outer Banks. The island's only highway was closed by waves and flooding. The producer told me that he wanted to pursue a new angle; instead of emphasizing human misery, the network wanted to cover the damage done by nature to nature. I enthusiastically agreed to be interviewed the next morning and even complimented the producer for having come up with a good idea. And with decades of observations behind me, I had all kinds of stories about storm damage on undeveloped barrier islands.

Like the time cooped-up storm waters in Pamlico Sound rushed catastrophically back to sea across Bodie Island, North Carolina, to form Oregon Inlet during an 1864 storm. Or the time that the same inlet expanded from a half mile to two miles in width during the course of a stormy 1962 evening. Or the complete disappearance of Mississippi and Louisiana islands in storms. Or the appearance of a brand-new barrier island after a storm off Florida's west coast. Even the flattening of island forests and the flying pine needles and sand that blinded numerous deer along the South Carolina coast by Hurricane Hugo in 1989 could have made great TV copy.

That afternoon and evening, however, even as Dennis roared ever closer to the mainland, my enthusiasm for the interview began to wane. I thought long and hard about "damage" to barrier islands. These long and narrow islands are very dynamic deposits of sand or gravel with the ocean on one side and a lagoon on the other that separates the island from the mainland. They make up 12 percent of the world's open ocean shorelines,

and they constantly change their shape and size and even their location. How could alterations caused by storms be called damage? Hurricanes are godsends for barrier islands.

Whether they strike the Outer Banks, or the far north's Arctic shoreline of Siberia, or a tropical island paradise in Brazil, storms provide the energy to shift sand to build the islands, to carve out new inlets, and to move the islands landward as sea level rises. Storms are the engines that drive barrier islands. No storms, no islands. The more I thought about it, the more I realized that the only hurricane damage on islands was to buildings, roads, and telephone poles that got in the way.

By the time the camera crew arrived I was ready. I gave what I thought was a well-thought-out argument that hurricanes not only didn't damage barrier islands, they were a critical part of the islands' evolution. Without hurricanes these islands would be swallowed up by the rising sea level. The camera crew told me I had done splendidly, but the producer didn't see it that way. Back in New York, my segment landed on the cutting room floor. I later heard that the interview was shown on some international airline flights where failed interviews are used to fill in dead time slots for the few passengers who are still awake.

I suppose it's easy to understand the attitude that something causing so much spectacular harm to human possessions must also damage nature. After

BATIK 1.1 *(opposite)*
Hurricane Season, 1998, 50″ x 36″
Hurricane Hugo approaches the South Carolina coast in 1989. A decade later Hurricane Dennis, a more poorly organized storm with a less distinct eye, followed a similar path but eventually went ashore on the Outer Banks of North Carolina.

Artist's note: On September 21, 1989, Hurricane Hugo struck Charleston, South Carolina, packing winds of 135 miles per hour, which destroyed both my studio and years of work. *Hurricane Season* is derived from a National Oceanic and Atmospheric Administration computer-generated image. The silk was commissioned by the H. John Heinz III Center for Science, Economics, and the Environment in Washington, D.C., and used for the cover of *The Hidden Costs of Coastal Hazards for Risk Management and Mitigation,* published in 1999 by Island Press.

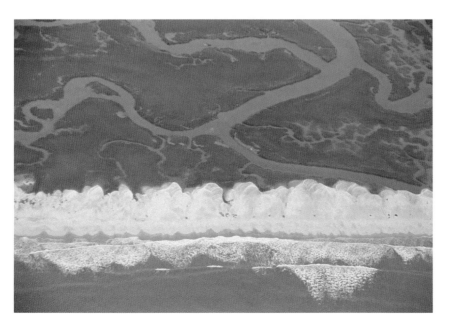

FIGURE 1.1
This photograph, taken on September 22, 1989, the day after the passage of Hurricane Hugo, shows barrier island migration in action. Storm waves caused the ocean side of this island (at Cape Romain, South Carolina) to retreat. The same waves widened the island as they washed sand across and into the lagoon (on top of the salt marsh). The result was landward movement, perhaps three or four feet, of the entire island. A large salt marsh with numerous meandering tidal channels is behind the island. *Photo by Rob Thieler.*

FIGURE 1.2
Some small South Carolina barrier islands, north of Edisto Island, barely wider than the beach. The islands are separated by inlets, each accompanied by a body of sand that protrudes out to sea, called the ebb tidal delta. The light-colored bands of sand covering salt marsh behind the islands are recent storm overwash deposits that widened the islands. *Photo by Mary Edna Fraser.*

all, an island without buildings has a much-changed appearance following a storm. But it is a big mistake to think that a storm is a natural disaster.

Buildings that line hundreds of miles of barrier island shorelines around the world are inflexible objects, protected only by the strength of concrete and steel. Barrier islands follow a different strategy. They respond to violent storms by bending, changing, moving and later recovering. And they incorporate these very processes into their strategy for existence. Barrier islands are among nature's most flexible and dynamic environments.

Dennis (late August and early September 1999) was larger than the average North Atlantic hurricane, but it wasn't one of the memorable ones, at least as many people measure them. The only person the storm killed directly was a middle-aged Florida surfer. Dennis didn't knock down a lot of houses either, or throw numbers of fishing vessels up on the beach. It did, however, move a lot of sand, and it proved to be a large step in the evolution of North Carolina's Outer Banks.

One of six hurricanes to strike the North Carolina coast in the final five years of the second millennium, Dennis's winds were much less powerful than those of Hurricane Fran (1996) and Hurricane Floyd (1999). Both were category 3 storms on the Saffir-Simpson Hurricane Intensity Scale (on a scale of 1 to 5, the higher the number, the greater the wind velocity of the storm) when they crossed the shoreline. During most of its lifespan Hurricane Dennis was a category 1 storm, but wind velocity is only one facet of storm character.

The big ones usually form off the west coast of Africa, but Hurricane Dennis originated a couple of hundred miles east of the Turks and Caicos Islands of the Southern Bahamas on August 23, 1999. By August 24 it was a tropical storm heading north, and two days later it was a hurricane. By August 28 the winds of Dennis reached their peak intensity of 105 miles per hour, or category 2. By then, Dennis was uprooting trees and power lines as it rolled over the Bahamas.

It maintained its category 2 status until August 30. By that time it was already pounding the coasts of North and South Carolina, having chased Vice President Al Gore from his Figure Eight Island, North Carolina, summer vacation and causing a shutdown of the Southport, North Carolina, nuclear power plant. More than twenty freighters caught in the storm path radioed wind velocities and barometric pressures to the National Hurricane Center in Miami. One can easily imagine the radio operator holding fast to a heaving, rocking table as a roaring, blinding, horizontal rain pounded the windows on the bridge and the occasional green water wave crashed over the bow. Then the storm began to move to the east-northeast, away from the coast. Headlines breathed sighs of relief; DENNIS SKIRTS THE COAST, DENNIS HEADS OUT TO SEA, and COASTAL RESIDENT'S PRAYERS ANSWERED.

Alas, the storm began to stall while the eye was 115 miles off the Outer Banks (batik 1.2). Soon it was wandering westward and southwestward,

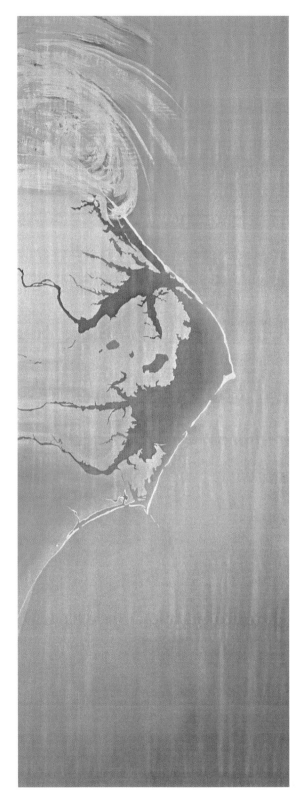

BATIK 1.2
Outer Banks, N.C., 1994, 152" x 55"
The sweeping arc of the barrier island chain of the Outer Banks of North Carolina. These are narrow and low islands, subject to high waves and low tides and are extremely dynamic, even by global barrier island standards.

Artist's note: Outer Banks is based on Landsat photographs of the North Carolina coastline provided by the U.S. Geological Survey in Reston, Virginia, with additional references to maps. The art also relates to breast cancer. The land configuration is vulnerable, and the passing storm pays homage to women who have suffered this life-altering threat.

slowly weakening, but all the while churning up the surface of the sea, pushing ever bigger waves out in all directions. Still a powerful tropical storm with winds of 70 miles per hour, Dennis continued rambling about in seemingly aimless fashion until, on September 4, it finally crossed the Outer Banks north of the Cape Lookout Lighthouse on Core Banks.

The storm had pounded the Outer Banks for almost a week, which was extraordinary behavior for a hurricane, much more like a winter storm or a nor'easter. Hurricanes usually pass over the shoreline quickly, but nor'easters tend to stay around and pound the beaches and thus generally move more sand on barrier islands than hurricanes do.

The worst storm to strike the Outer Banks in the twentieth century was the 1962 Ash Wednesday storm, a nor'easter that lingered off the East Coast for three full days, smashing beachfront buildings on barrier islands from Massachusetts to north Florida. This great storm struck during the twice-a-month *spring tides*, which are higher than normal tides because of the optimal alignment of the sun and moon. Extra-high tides bring the storm waves further onto the islands, leading to more overwash and flooding. The great eastern North Atlantic storm of 1953 that breached the dikes of Holland also struck during the spring tides.

Dennis brought as much as twenty inches of rain to the barrier islands, and the torrents that fell on the mainland as Dennis crossed the shoreline set the stage for the monumental flooding disaster that occurred inland. Some of the coastal plain streams were still in flood stage from Dennis when the even more rain-laden Floyd struck a week later.

Three things determine the size of storm waves: the *intensity* of the winds; the *fetch*, or open-water distance over which the wind blows; and the *duration* of the storm. The more intense the winds, the greater the fetch, and the longer the duration, the larger the waves. Two things that Hurricane Dennis had plenty of were fetch and duration.

It was a stroke of good luck that Dennis struck the United States Army Corps of Engineers wave research pier in Duck, North Carolina, a facility that was just waiting for such an event. Measurements made during Dennis indicate that the storm *wave height* (the vertical distance between the wave trough and the wave crest) exceeded twenty feet. The wave runup on the beach (the highest reach of the waves) was close to the greatest on local record. The wave period (the time required for the passage of successive wave crests) was thirteen or fourteen seconds. A big storm in the Arctic where the fetch is small because of the ice pack might have a wave period of three seconds. Normal waves on a breezy summer day on the Outer

Banks might have periods of seven to ten seconds. The 1991 Halloween storm, where the waves formed far offshore, had waves periods of twenty-three seconds, but this giant of giants (the perfect storm of the book and the movie) caused fewer changes to the barrier islands than Dennis.

Bill Berkemeier, the director of the Duck Pier facility, points out that because of Dennis's unusually long duration, the hurricane expended an unusually large amount of energy on the Outer Banks. It was one of the most significant storms during the twenty-year existence of the pier, as measured by work accomplished in moving beach, island, and shoreface sand. Dennis produced more extensive overwash on Core Banks and Portsmouth Island than I have seen in thirty-five years of Core Banks watching.

LIVING WITH NATURE ON THE OUTER BANKS

Of the world's twenty-two hundred barrier islands, the five constituting North Carolina's Outer Banks are among the most renowned (fig. 1.3). The thin line of sand, far from the mainland and extending on either side of Cape Hatteras, is a familiar part of the United States East Coast outline. Site of many shipwrecks, famous lighthouses, wild horses, and legendary storms, the Outer Banks achieved a new level of recognition in 1999 when the 3,500-ton Cape Hatteras Lighthouse was moved away from an eroding shoreline, in the center of an international media spotlight.

From a global perspective, the Outer Banks are more dynamic than most barrier islands. Waves are large because the continental shelf is narrow there and doesn't provide as much wave-dampening friction as would a wide, flat shelf like that along the Gulf of Mexico. The islands, in turn, are mostly narrow and are frequently overwashed by waves. Perhaps because things happen so quickly, these were the first barrier islands to be studied in detail.

Much of the Outer Banks is within the confines of two national seashores (equivalents of national parks). The Cape Hatteras National Seashore extends from just south of Nags Head to Ocracoke Inlet, encompassing a part of the Currituck Peninsula and Hatteras and Ocracoke Islands, coexisting somewhat uncomfortably with eight small towns. Cape Lookout National Seashore extends from Ocracoke Inlet to Beaufort Inlet and includes three islands—Portsmouth, Core, and Shackleford Banks. These barrier islands are among the most pristine in North America, with no beach cottage communities interrupting the horizon.

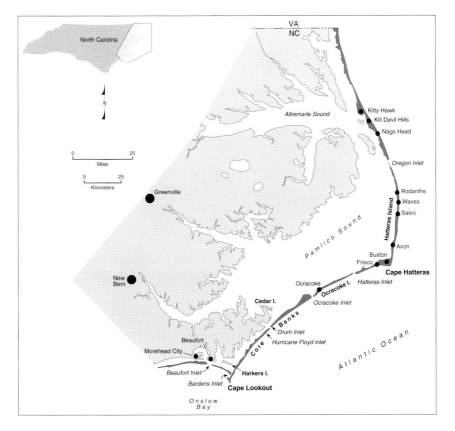

FIGURE 1.3
Map of the Outer Banks of North Carolina.

By the time Hurricane Dennis struck in 1999, the Outer Banks from Oc-racoke north had been much altered by humans. The most profound change was introduced in the 1930s, when the islands were occupied mostly by fishing villages with almost no beachfront development. Everyone knew that the barrier islands were eroding; in fact, they were eroding on both sides. It seemed they would soon disappear. As luck would have it, the country was in the middle of the Great Depression, and the federal government eagerly sought projects that would require lots of manpower. What better scheme than to build a large sand dune down the length of the islands (fig. 1.4), "saving" them from erosion while simultaneously providing thousands of jobs?

The Civilian Conservation Corps (CCC) and the Works Progress Administration (WPA), two make-work agencies created by the desperate times, constructed the dune. The dune construction workers, especially the young men of the CCC, were not always happy with their isolated existence on the Outer Banks. To keep them on the job, the CCC instituted a Depression-era rule that anyone who went AWOL would not be allowed to join the army, a rule that was quietly dropped at the commencement of World War II.

FIGURE 1.4
Construction of the Civilian Conservation Corps dune in the 1930s, using brush and driftwood to make sand fences. The Cape Hatteras Lighthouse is in the background. *Courtesy of the Outer Banks History Center and the National Park Service, Manteo, N.C.*

The problem with the new-dune idea was that it was founded on the poor assumption that the Outer Banks really were eroding away. Actually, they were either migrating or thinning in preparation for migration. Unknown in the 1930s, barrier island migration was an idea whose time wouldn't come until the early 1970s. As it turns out, not only was the idea of dune construction ill founded, it actually exacerbated the problem it was intended to solve. The dune caused the rate of shoreline retreat to increase.

Before the construction of the big dune with bulldozers and mules, drag pans and sand fencing, the seaward-most one hundred to two hundred yards of much of the Outer Banks was a broad, flat, storm overwash band (fig. 1.5), lightly vegetated and frequently traversed by storm waves that brought in fresh sand. It was also critical habitat for shorebird nesting, habitat that disappeared almost in its entirety when the dune was constructed.

FIGURE 1.5
Before the Civilian Conservation Corps
dune was built, storm waves rolled across
this section of Nags Head frequently
enough to prevent establishment of a
vegetation cover. The row of beachfront
houses shown here in June 1936 is the first
such development on the Outer Banks.
*Courtesy of the Outer Banks History
Center and the National Park Service,
Manteo, N.C.*

After dune construction, the islands began to respond to storms in an entirely different way. No longer could storm waves carry sand into the islands. The new dunes were so high that in most cases they prevented even windblown sand from coming onshore off the beach. When storms came, all the energy of the crashing waves was now expended on the beach. No longer was breaking-wave energy partially absorbed as the water rolled across the islands. Assault of the beach with all that wave energy led to increased rates of shoreline erosion. Bushes and a few trees that once were restricted to a narrow strip along the back side of the islands began to creep forward to the front side, since the dunes protected them from wind and salt spray. The broad, light-colored band of unvegetated overwash sand next to the beach was gone. Now it was "safe" to build buildings and highways next to the beach in the shelter of the dune. The rush to the beach began.

The highway didn't come until after World War II. Before that, towns on the barrier islands were connected by numerous wandering sand ruts (fig. 1.6) linked by ferries at Oregon and Hatteras Inlets. Old-timers claim that there were 108 roads on Hatteras Island alone, and when strangers got off the Oregon Inlet Ferry and asked for instructions on how to get to the cape, they were instructed to take road 108! Today a single road, North Carolina State Highway 12, runs the length of the Outer Banks from Ocracoke to Corolla. Highway 12 even continues on the mainland for a few miles, connected by the Ocracoke Ferry across Pamlico Sound.

As the blacktop road neared completion in the early 1950s, a problem arose. Few on the islands had a driver's license, and driving on the new blacktop road required one. The state sent a crew of examiners to give the driver's license test to five hundred Outer Bankers simultaneously. Most flunked initially, since few had any mainland experience with signs, stoplights, and turn signals.

The road was constructed mostly just behind the big artificial dune and occasionally, as the shoreline predictably retreated, sections of the road fell in. After a while the formerly straight road acquired a few corners and curves, each marking a bypass around the site of a "going-to-the-sea" highway segment. The dune thinned over the years, and beginning with the giant 1991 Halloween storm, hundreds of new dune breaks appeared, too many to repair. Highway 12 was in real trouble (fig. 12.19). Hurricane Dennis would further up the ante. It interacted with islands in both national seashores in three distinctly different fashions; two favorable and one unfavorable for the island's futures.

Dennis and Core Banks

The response of Core Banks (batiks 1.3 and 1.4) to Hurricane Dennis offers a rare opportunity to view the impact of a storm on a low and narrow (one hundred yards) island, unfettered by human paraphernalia. The CCC Dune did not extend to Core Banks. During the storm's weeklong assault, winds with widely varying velocities came from every direction, depending on where the storm was located at a particular time. For days on end, the surf roared, the air remained thick, hazy, and salty, and the waves moved up and down, back and forth forever.

The first thing the storm did was to flatten the beach by removing sand from the upper reaches and from the dunes and depositing it on the lower parts of the beach. The broad, flattened beach paved the way for waves to roll into the island, carrying and depositing sand in tongue-shaped overwash

BATIK 1.3
Core Banks, N.C., 1995, 48″ x 35″
Core Banks, shown here, is part of the Cape Lookout National Seashore and is a rare natural laboratory to study barrier island migration. During Hurricane Dennis, the island migrated a foot or so toward the mainland.

Artist's note: Core Banks is an aerial view with colors inspired by woodblock prints from Japan's Edo period (1615–1868). My father, Claude Burkhead Jr., piloted our 1946 Ercoupe to the Outer Banks of North Carolina on a mission to supply Pilkey with photographs in February of 1995. My father is a Certified Flight Instructor (Instrument), and we flew in 35 mph winds that day. Cape Lookout Lighthouse anchors this design.

BATIK 1.4

South of Ocracoke, N.C., 2001, 36″ x 36″
The north end of Core Banks, sometimes called Portsmouth Island, is in the foreground, and Ocracoke Island is at the top. This end of Core Banks is very wide, and the waves from Hurricane Dennis penetrated perhaps halfway across the island. The storm protection afforded by the width is why the small nineteenth-century fishing village of Portsmouth was located here, just off the batik to the left.

Artist's note: Flying over the Outer Banks, my father and I were looking for the signs of overwash as one of my aerial assignments for Pilkey. This, along with the lobes on the back side of the island, made for a dynamic batik. In a small plane you feel the fragility of these barrier islands. You also recognize the intrinsic connection between the forces of nature and the quiet rapture of a walk on the beach.

fans (fig. 1.7). Most of the overwash sand penetrated only a few tens of yards, but the occasional large wave combined with a high tide and high storm surge level to push sand completely across the island and into the sound behind. The net result was an island with a slightly higher average elevation than before the storm.

What happened on Core Banks was a tiny step in the island's long-term response to sea level rise. Because the ocean-facing beach retreated a bit and the mainland-facing lagoon side of the island built out a bit into the lagoon, the entire island moved landward ever so slightly. Hurricane Dennis caused the island to migrate!

Dennis and Shackleford Banks
The western half of Shackleford Banks responded much differently. This part of Shackleford is a high and wide, east-to-west-oriented island. Dune

FIGURE 1.7

Overwash fans from Hurricane Dennis on Core Banks. The white fingers of sand are fresh sand bodies that eventually will merge with the island as buried vegetation grows through the sand. Note that most overwash sand was deposited near the beach, a process that, over time, makes the beachfront the highest point of elevation on the island. *Photo by Andy Coburn.*

heights commonly exceed twenty feet, affording protection for more than 280 species of plants. In contrast, the low and thin Core Banks has only 35 species of plants, all hardy and able to withstand the constant onslaught of wind and salt spray with occasional flooding.

Along this stretch of shoreline, overwash during Dennis was unimportant because it could penetrate the front dune line in only a few places. Instead, the front dune row eroded, leaving a continuous cliff or *erosion scarp*, six to ten feet high, along the beach. Simultaneously the backbarrier side of the island eroded, as evidenced by newly fallen trees on the lagoon beach (fig. 1.8). Shackleford Banks grew narrower as a result of Hurricane Dennis. Probably the combined result of Dennis and Floyd was thinning of the half-mile-wide island by ten to fifteen feet.

While Shackleford Banks is not migrating, it is preparing for migration by thinning down to the point that cross-island overwash can readily occur, as it did on Core Banks. The narrowing on Shackleford can be viewed as a logical response to the sea level rise. In fact, all of the Outer Banks that are not already thinned down to one hundred to two hundred yards in width are getting thinner in preparation for migration. At least, that's my interpretation.

One of the old cemeteries in Rodanthe, a village in the Hatteras Seashore, offers unique proof of island thinning. The parishioners always placed their tombstones facing the road, the direction from which visitors

would approach. All the pre-1940 tombstones now face the sound, while those after the 1940s face Highway 12 in a seaward direction. That's because the local pre-1940 and pre–Highway 12 road fell victim to shoreline erosion and island narrowing and is now underwater out in Pamlico Sound.

Dennis and the CCC Dune

The rest of the Hurricane Dennis story is a tale of the CCC Dune that, by its presence or absence, controlled the islands' response to the storm. Overwash occurred where the dune had been breached by past storms. Where the dune survived or had been repaired by the North Carolina Department of Transportation, the waves expended themselves in the surf zone and change was limited to dune erosion. After the passage of Dennis, spectacular vertical dune erosion scarps as high as twenty feet were visible to those who cared to hike to the beach.

Hatteras and Ocracoke Islands, "protected" by the CCC Dune, were not allowed to migrate like Core Banks during Dennis. Nor did they respond like Shackleford Banks, in such a way as to prepare for future migration. The encounters of Hatteras and Ocracoke with Hurricane Dennis were mostly futile attempts to throw off the human yoke and bring the islands back to a natural state.

South of Nags Head, there were ten "hot spots," so designated by the North Carolina Department of Transportation, where overwash closed

Highway 12 by burying it in sand. Overwash events that were beautiful examples of an evolving, migrating barrier island on Core Banks became part of a "natural disaster" on the Cape Hatteras National Seashore. The island elevation gained by the deposition of overwash sand was quickly negated by road crews that removed the sand from Highway 12 and bulldozed it back to the beach. Just north of Buxton, within sight of the newly moved Cape Hatteras Lighthouse, waves crashing through the artificial dunes tore up a full half mile of the asphalt and carried it into the lagoon. The washout occurred exactly at the site of an inlet that formed in the 1962 storm (later to be filled in artificially).

The main channel of Oregon Inlet, named for the first vessel to sail through it, deepened as returning storm waters rushed out of Pamlico Sound when the storm passed inland. The state transportation department held its breath. If the storm blew out the inlet channel like the 1962 storm did before the Bonner Bridge was built over it, the bridge would collapse. The pilings were far too short to withstand another storm "blowout."

Because Dennis hung around for so long, it was possible to slightly repair dune gaps (fig. 1.9) between each high tide. All up and down the banks, especially in the villages, an army of graders, bulldozers, front end

8/31/1999 12:19pm

FIGURE 1.9
Here it comes! During Hurricane Dennis, a wave begins to roll into the island down an overwash alley formed by two houses. *Photo by Andy Coburn.*

FIGURE 1.10
The Comfort Inn at South Nags Head, This building, the first high-rise on the Outer Banks, was built well back from the beach. Now its time is almost here. First, more than a decade ago, a restaurant with a magnificent view of the sea fell in. Next, the swimming pool was destroyed by Dennis, and a future storm will most likely destroy the whole building. *Photo by Andy Coburn.*

loaders, and anything else that moved and carried sand suddenly appeared at each low tide and for a few hours frantically worked against the islands' attempts to build up their elevations. The heavy equipment pushed the latest layer of overwash sand into a ridge on the upper beach, only for it to be destroyed in the next high tide as the process began anew. It was sad to witness—Mother Nature losing a battle. But perhaps she has not conceded the war (fig. 1.10).

The final disastrous blow, struck by Hurricane Dennis on September 4, was to turn my retirement party in Hillsborough, North Carolina, 180 miles from the sea, into a deluge!

The Global Picture

N URTURED by sand and forever shaped by winds, waves, ice, and changing sea levels, barrier islands endure and persist. Capable of withstanding the largest of storms, they require natural catastrophes for their very survival. They can build upward if earthquakes or other forces cause the land to sink, and most striking of all, they are capable of moving back toward the mainland when a rise in the level of the sea threatens to consume them. Barrier islands exist all over the world, in all climates, along all types of coasts (batik 2.1) and yet are among the youngest major landforms on the surface of the earth. Few are more than four thousand years old, and some are only a few months old.

As each island is unique, each evolves in an exclusive combination of physical, biological, atmospheric, and oceanographic circumstances. The uniqueness of the islands was demonstrated to me in a most painful way on my first visit to the Algarve, along the south coast of Portugal. The Portuguese Geological Survey wished to start a coastal geology program, and I was asked to introduce a group of Portuguese marine geologists to the vagaries of coastal or nearshore marine geology. On the first stop of a two-week field trip along the length of the Portuguese shoreline, we stood on a barrier island beach as I confidently explained that the black shells scattered about the beach were certain evidence of island migration. Based on my experiences in North Carolina, I indicated that the shells became stained after the death of the animal and the burial of the shell. The color came from iron sulfide precipitated in microscopic cavities in the shells after they were buried in lagoon mud located behind

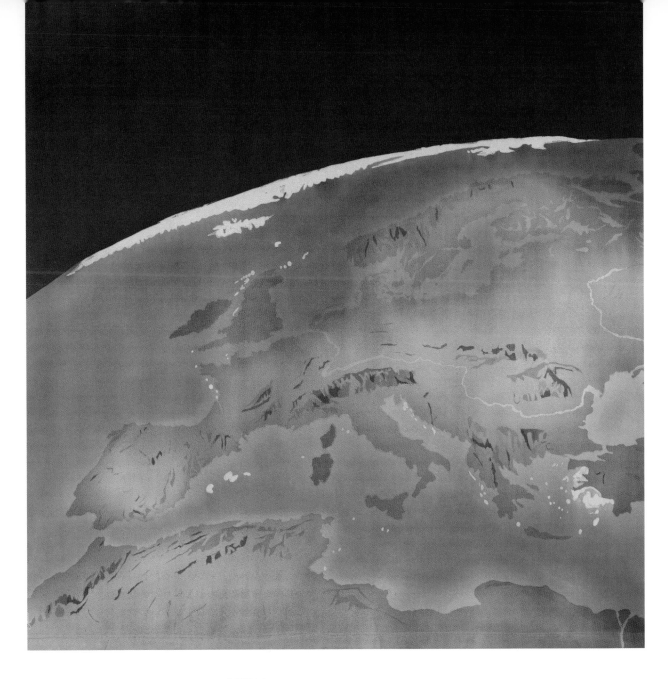

BATIK 2.1

Earthscaping, 2000, 47" x 47"

A global view of Europe and the rims of North Africa and Greenland. Chains of barrier islands within this vista include South Portugal, the Nile Delta in the Mediterranean, and Holland, Germany, and Denmark facing the North Sea. In addition, a few islands rim the Arctic Ocean. Relative to North America, the number of barrier island chains in Europe and North Africa is small.

Artist's note: This global simulation is an artistic rendition of a computerized ad in a magazine. The Earthscape features the North Atlantic and the Mediterranean. I found the natural curve of our world and the oblique topographic features to be intriguing. The indigo sky complements the mustard-colored land.

the island. When the island migrated over the shells, they were exposed anew on the beach.

When I finished the narration, I suggested that we next dig a small ditch across the beach to observe and discuss the layers in the sand. Two spadefuls later we encountered a stinking mass of rotting seaweed, within which all the shells had been stained black. I had never seen buried seaweed on a beach before. Clearly black shells on a Portuguese barrier island beach could not be used as a signal of island migration. The shells had been stained black in the beach, not earlier in the lagoon. My confident recital on the significance of black shells in beach sand worked perfectly for North Carolina barrier islands but was dead wrong for Portuguese islands.

My embarrassment evaporated as the exciting trip unfolded, but it was a lesson on barrier island uniqueness that I never forgot. Each barrier island is different. Whatever one learns about the evolution of one island, even relative to such minor aspects as the color of seashells on the beach, may not apply to another.

GLOBAL DISTRIBUTION

Barrier islands, nearly twenty-two hundred all told, are found on the edges of every continent except Antarctica. Small barrier island chains even exist in large lakes. Somewhere around 12 percent of all the open-ocean coasts of the world are fronted by these islands (appendix, tables 1, 2, and 3). More barrier islands (73 percent) exist in the Northern Hemisphere than in the Southern (27 percent), mostly because there is a lot less land area and a lot less shoreline south of the equator. The difference also reflects differing histories of sea level change between the Northern and Southern Hemispheres in the last few thousand years.

The country with the largest number of barrier islands (405) and the greatest length of barrier shoreline (3,054 miles) is the United States. Russia is a distant second with 226 islands that extend over a shoreline distance of 1,020 miles. About a fourth of all barrier islands are "American-owned," one reason why much of the world's scientific literature on barrier islands is of U.S. origin.

On a larger scale, global barrier island distribution is closely related to the nature of the mobile tectonic plates that make up Earth's crust. Our planet's outer shell consists of eight major plates and numerous smaller ones. These plates move as a result of the upward transfer of heat from deep within Earth. A plate with a continent sitting atop it is called a continental plate. The continents "float" above the ocean floors because they are made of lighter, granitic-type rocks. Plates underlying oceans are thinner and heavier, consisting mostly of basaltic rock (a type of lava). Where a continental plate and an oceanic plate collide (converge), as on the Pacific coasts of North and South America (batik 2.2), mountains line the narrow, steep coast. Only 15 percent of the barrier islands lie along these shorelines, most of them on the rims of sand-rich river deltas such as the Copper River Delta in Alaska and the Río Patia Delta on the Pacific coast of Colombia. There, swiftly flowing rivers carry sand and mud to the sea to form a platform that provides the requisite flat surface along the outer edge of which barrier islands form.

On the "quiet" side of the moving continents (batik 2.3), the oceanic plates and continental plates move as one, and flat coastal plains form as the continents shed their detritus into the ocean basins over millions of years. Seventy-three percent of barrier islands are found on the quiet sides of continents, including the east coasts of North and South America.

Chains—three or more adjacent islands or sometimes upwards of thirty in a group—form along both coastal plains and river deltas. The longest barrier island chain in the world is in northern Brazil: fifty islands along 356 miles of shoreline. Next in length is the chain on the rim of the Niger River Delta in Nigeria, which consists of thirty-eight islands, along 343 miles of shoreline. Barriers adjacent to coastal plains are found mostly in the Northern Hemisphere, along the shores of the Arctic Ocean (fig. 2.1) and along the North American East and Gulf Coasts. In the Southern Hemisphere, most barrier islands are formed on the rims of deltas. Examples include the Gurupi islands of Brazil and the East African islands off Madagascar. Single islands, such as Matakana Island in New Zealand, Iztuzu Beach in Turkey, and Bartagh Island in Ireland are most likely to be formed across bays or river mouths on steep, rocky coasts.

More than plate tectonics and flat surfaces is involved in determining global barrier island distribution. For example, why are no barrier islands found along the coastal plain where the Amazon River flows into the sea? Why are there none across the mouth of the Bay of Fundy between New Brunswick and Nova Scotia, Canada? Why are there no barrier islands along Florida's big bend coast in the northeast Gulf of Mexico? And why none at Myrtle Beach, South Carolina?

It turns out that coastal plain barrier islands have five requirements:

- a rising sea level
- a gently sloping mainland surface
- a supply of sand
- energetic waves
- a low to intermediate tide range

When these elements are present, there will be barrier islands. When one or more of the elements is absent, no barrier islands will occur.

The shoreline off the Amazon River is furnished with a huge supply of sediment, the largest of any shoreline in the world, but only a small portion of the sediment is sand. Most of it is mud, which does not pile up into ridges like barrier islands. Instead the shores of northern Brazil are lined with broad mudflats that seem to extend to the horizon. The big bend coast of Florida, between the Florida panhandle and the peninsula, is a long stretch of marshy coastline adjacent to one of the widest (130 miles) U.S. continental shelves. A plentiful sand supply exists offshore, but the shelf is so wide and flat that the energy of the waves is sapped by friction with the seafloor long before they reach the shoreline. They simply lack sufficient energy to move sand grains at the shoreline and shape them into islands. Barrier islands do not form across the mouth of the Bay of Fundy because roaring tidal currents quickly wash them away. The range between high

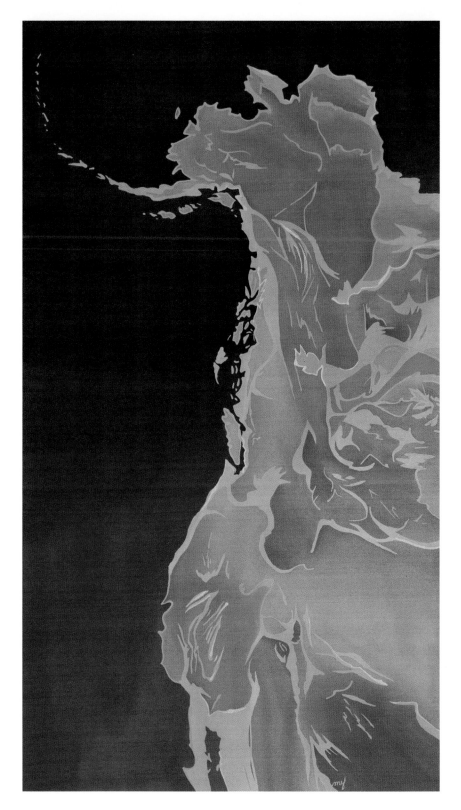

BATIK 2.2
West Coast USA, 1995, 84″ x 48″
The West Coast of North America is a steep, tectonically active zone, with relatively few barrier islands lining the shore. There are barrier islands along the central Baja California coast of Mexico, and along several stretches of the Alaska coast where deltas form, south of the Aleutian Islands. In the far north, a large number of barrier islands are found along Alaska's Chukchi and Beaufort Seas.

Artist's note: Batiks 2.2 and 2.3 were commissioned for the United States Embassy in Thailand. They hang on each side of the elevator when you go to see the ambassador in Bangkok. *West Coast USA* was derived from a map in the *New York Times Atlas of the World* and *East Coast USA* from a NOAA, AVHRR Mosaic in *Looking at Earth*.

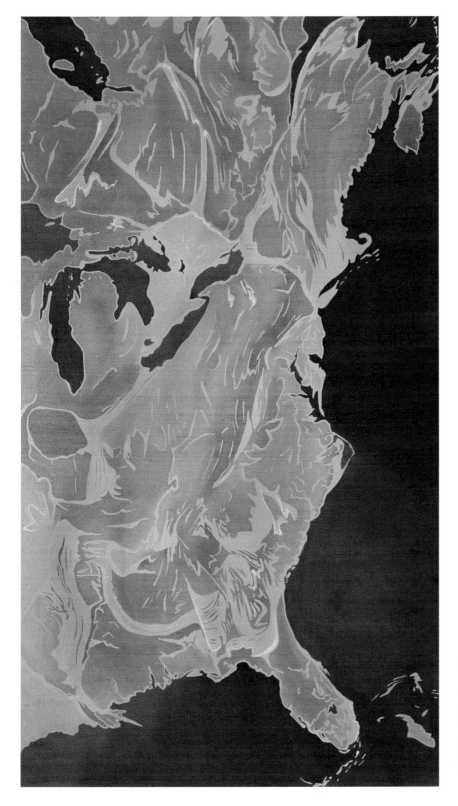

BATIK 2.3
East Coast USA, 1995, 84″ x 48″
There are more barrier islands along this
coast than along any other in the world.
The islands form an almost continuous
chain from New York to Mexico. There are
a number of individual barrier islands and
small island chains along the coasts of
New England and the Maritime Provinces.

and low tide can be as great as forty-eight feet, and few barrier islands can coexist with tide ranges greater than fourteen feet.

The Myrtle Beach coast is a different matter. All the requirements for barrier islands are present, but the islands aren't. A few thousand years ago, the migrating barrier island chain merged with the mainland, combining its sand volume with the much larger Ice Age island left stranded on the coastal plain by a higher sea level, probably 120,000 years ago. This lack of a barrier island is only temporary, however. As sea level continues to rise, the Ice Age islands should be surmounted or eroded away, and a new barrier island chain will then form.

Delta barrier islands do not depend on sea level rise for their origin. They do not require a flat coastal plain to form, either. And the fourteen-foot maximum tide range for coastal plain islands does not apply. The Gurupi Delta barrier islands in Brazil coexist with twenty-two-foot tides. These islands usually form as waves chew into the seaward-extending lobes of river delta sediment. Easily suspended mud is dispersed, leaving behind sand to form a beach and an island. The requirements for their formation are

- a river delta platform
- a supply of sand
- energetic waves

VARIETY IS THE SPICE OF LIFE

Despite the mélange of island settings and characteristics, there are common threads. Like most things natural, however, barrier islands defy precise global classification. Nature never totally cooperates with those who want to conveniently categorize its attributes. Nevertheless, for our purposes, categorizing boundaries between island types can be very educational, even though imperfect.

Most fall into three broad categories: coastal plain, delta, and Arctic. If, however, we include some relatively uncommon types, subdivisions of the world's barrier islands expand to nine:

- **coastal plain barrier islands**, the most common variety, exemplified by the island chains on the Atlantic and Gulf of Mexico coasts of the United States. Example: the Outer Banks of North Carolina
- **delta barrier islands**, fringing the edges of the lobes of mud and sand extending seaward from river mouths. Example: Timbalier Island, Louisiana

- **Arctic barrier islands**, a special case of coastal plain barrier in which ice plays an important role in island evolution. Example: Cross Island, Alaska
- **bay mouth barrier islands**, usually occurring singly and enclosing any type of small coastal embayment along mountainous shorelines. Example: Matakana Island, New Zealand
- **sandur barrier islands**, formed at the seaward edges of sediment fans (sandurs) in front of melting glaciers that supply a huge volume of sand to the shoreline. Example: southeastern Iceland coast
- **composite barrier islands**, Holocene (modern) islands combined with a preexisting stranded Pleistocene (Ice Age) coastal plain island. Example: the Georgia Sea Islands
- **accidental barrier islands**, rare eroded piles of sand (e.g., a coastal dune field) that did not accumulate on a barrier island but are now surrounded by water. Example: Fraser Island, Queensland, Australia
- **man-made barrier islands**, usually formed as channels are dredged behind and through ridges of sand bordering the shoreline. Example: Palm Beach, Florida
- **lagoon barrier islands**, found behind open-ocean islands. Example: Cedar Island, North Carolina

HOW AND WHY IT ALL BEGAN

Nature is at odds with itself. Rising sea level floods river valleys to create irregular shorelines, made jagged by endless series of bays, each extending up the former valleys. Crooked shorelines, however, are unbalanced, out of equilibrium. Protruding headlands become points of rapid erosion that provide sediment that is deposited as spits across the mouths of the bays (fig. 2.2). The whole process implies a return to equilibrium or a stable shoreline, where erosion balances deposition. Nature strives to undo the "damage" of a rising sea by constructing a straight shoreline, and barrier islands do just that. This is the fundamental reason why coastal plain barrier islands form.

If sea level is rising against a steep coast, as the western coasts of North and South America, for example, inundation of the river valleys is slight. There is no crooked shoreline to straighten, and few barrier islands form.

Most barrier islands formed seaward of where they are today, so the puzzle of island origin cannot be solved by studying today's islands. Instead, island origin must be inferred from our knowledge of how the sea works

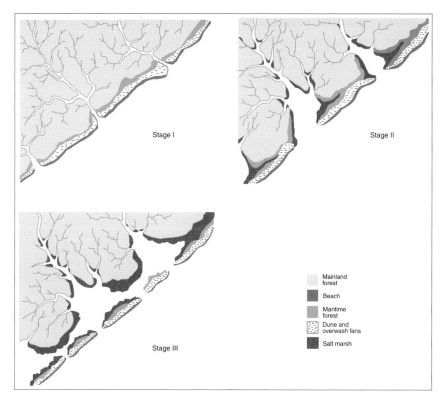

at the shoreline, how sand is moved by waves and wind, and what storms
do to islands. We have to extrapolate from known processes, as well as examine a few barrier island chains, like those in Holland, that formed within historic time.

The evolution of our understanding of barrier islands is a fascinating
study in itself. Hypotheses of island genesis, beginning with French geologist Elie de Beaumont's 1845 study, were proposed and then enthusiastically (even fiercely) defended by proponents. Coastal scientists aligned themselves behind their favorite explanations, assuming that one hypothesis
would apply to all islands and that any other ideas were wrong. Since field
data and evidence were scarce in the early days of the debate, the choice of
explanation was unencumbered by facts. Acceptance or denial of the various hypotheses was based largely on personal experience along a particular
shoreline segment somewhere in the world. As it turned out, just about all
the hypotheses were operational theories *somewhere*. In other words, everyone was right. As we shall see, islands were formed in many different ways.
Three main theories of origin of coastal plain islands emerged:

THE GLOBAL PICTURE

- **Spit breaching**. Sand spits form across the mouths of bays on a coastal plain recently flooded by a sea level rise. The spits are breached during storms to become islands.
- **Beach ridge isolation**. Ridges of dune sand, immediately adjacent to the beaches, offer some resistance to erosion, which allows the dune ridge to persist as sea level rises. The land behind the ridge is flooded, and the ridge breaks from the mainland to form a barrier island.
- **Submarine bar upgrowth**. A sandbar constructed during a storm accumulates enough sand to build above normal sea level. An island is left high (and dry) as storm waters recede.

The idea that barrier islands were formed when storms breached spits was proposed in the 1880s by G. K. Gilbert. Amazingly, Gilbert arrived at this idea while studying ancient spits formed during high lake levels of the ancestral Great Salt Lake in the desert of Utah. Later, in the 1960s, John Fisher carried out studies on the North Carolina Outer Banks that supported Gilbert's hypothesis. Fisher's brilliant doctoral dissertation at the University of North Carolina, although far ahead of its time, was never published.

Spits form across the mouths of bays and lagoons when the open-ocean waves encounter the bay and are forced to bend or refract in order to conform to the bay shoreline. In the process of bending, the waves lose energy, which in turn results in loss of their sand-carrying capacity. Therefore, sand is not carried up the bay and deposited along the bay shoreline. Instead the sand is dropped right at the entrance to the bay, and spits are built out into deeper water straight across the former river valleys. Inevitably, storms break through the spits and form islands.

When we lived at the University of Georgia Marine Institute on Sapelo Island (a barrier island), John Hoyt, our next-door neighbor, argued that islands formed when sand dune beach ridges were flooded from behind and isolated by a rising sea level. Hoyt's idea was a modification of de Beaumont's hypothesis. Hoyt's work on barrier islands in the 1960s helped revitalize the study of these features at the time when the great island real estate boom was beginning in the southern United States. His premature death in a glider accident was a great loss to barrier island research.

Ervin Otvos, a University of Mississippi geologist, has been the chief proponent of "offshore bar buildup" origin for islands in recent years. This mode of island origin is probably an uncommon event, restricted to very flat and broad continental shelves with normally low waves. During the high storm surges that sometimes occur on such continental shelves, sandbar heights build up in equilibrium with a high sea level. After a storm, quick lowering of the water level leaves a bar intact and above sea level. Anclote Key and Dog Island along the West Florida big bend coast are believed to have formed by such submarine bar upgrowth.

Perhaps the individual most responsible for our current understanding of barrier island evolution is coastal geologist Miles Hayes, formerly with the University of South Carolina and now an expert on oil spill cleanup. Besides initiating studies of many individual barrier islands all over the world, Hayes recognized the intertwined roles of the vertical range between high and low tide levels and wave height and also of the tidal deltas in island development. If one considers the continuing contributions of his students and associates, including Al Hine, Jon Boothroyd, Dag Nummedal, Dennis Hubbard, Duncan Fitzgerald, Larry McCormick, Tom Moslow, and Victor Goldsmith, it becomes clear that his legacy is tremendous.

ISLAND EVOLUTION

Once islands form, they begin evolving in many ways. They change size, shape, appearance, and even migrate back toward the mainland.

The Role of Sand Supply

Sand is like food for barrier islands. This nourishment comes from four principal sources: rivers, the shoreface or inner continental shelf, eroding shoreline bluffs or dunes, and sand introduced by longshore current transport along the beach from some source on adjacent shorelines. River and continental shelf sand is first supplied to the shoreface, or tidal deltas, and is then washed ashore to barrier island beaches by the action of waves and currents.

Other things being equal, a large sand supply makes for a large island. Islands grow wide by expanding (not migrating) in a seaward direction. Such islands (Bogue Banks, North Carolina; Santa Barbara Island, Colombia; and the Niger Delta islands in Nigeria) are called *regressive islands* (fig. 2.3). Each regressive island is characterized by a succession of dune ridges on its surface that parallel the shoreline, each representing a former shoreline during the widening process (fig. 2.4). Barrier island widening was first recognized by Shell Oil geologists who studied Galveston Island, Texas, to understand ancient, deeply buried barrier islands that might contain oil.

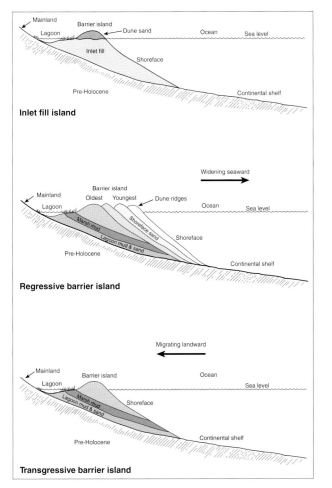

Inlet fill island

Regressive barrier island

Transgressive barrier island

FIGURE 2.3

Idealized line drawings with much vertical exaggeration, showing regressive, transgressive, and inlet fill islands. A regressive island is one that is widening in a seaward direction as new sand is placed on the shoreface and it builds out. A regressive island tends to be wide, and its surface expression will be a series of sand dune ridges (beach ridges), each of which represents a former shoreline position (see fig. 2.4). A transgressive island is one that is migrating or has recently migrated in a landward direction over old lagoon deposits. Such islands are low and narrow (see fig. 2.5). An island segment that has recently been occupied by an inlet will consist of sand that filled the inlet as it migrated to a new location. Inlet fill usually is beach sand, brought to the inlet by longshore currents.

FIGURE 2.4

A regressive barrier island on the Pacific Coast of southern Baja California. The beach ridges, characteristic of an island that is widening in a seaward direction, are extraordinarily well developed on this island. Each ridge represents a former shoreline position. *Photo by Miles Hayes.*

Regressive islands have a characteristic cross section of strata that reflects their history of widening.

If the sand supply is too large, islands can't form. For example, Dutch geologists believe that a large amount of sand from the Rhine River is responsible for the lack of barrier islands south of the Frisian Islands where the sand supply would fill in any lagoons behind any islands if they did form.

A low sand supply to the coast results in skinny islands such as the Outer Banks of North Carolina. Most thin coastal plain barrier islands are *transgressive islands* (fig. 2.3), or islands that are migrating toward the mainland (fig. 2.5). They are also the ones most likely to have lagoonal mud and shells exposed on their open ocean beaches.

The relative importance of sand sources is highly variable in time and space, from season to season, and from island to island. Adjacent islands may be supplied with radically different volumes of sand. For example, Core Banks, North Carolina, is between one hundred and two hundred yards wide, with few dunes; compare that to the half-mile width and high dunes of the adjacent Shackleford Banks. The difference in island orientation relative to the prevailing winds may be responsible. Shackleford is oriented east-west, such that onshore winds bring in large amounts of dune sand, but the same winds blow down the length of the north-south-oriented Core Banks and deliver much less sand from the surf zone to the island. During the most recent geologic history of Shackleford Banks, an inlet has migrated almost the entire length of the island, resulting in a cross section made up entirely of sand that filled the inlet (fig. 2.3).

The ancient harbor of Kuanos in Turkey, founded in the ninth century B.C., offers an example of the importance of sand supply (as well as the importance of human activities). When the countryside was deforested, hillside erosion was enhanced and the greatly increased supply of river sediment built up Iztuzu Island across the harbor entrance, forever closing the harbor to all but small vessels.

The Role of Sea Level Change
During the last three million years, sea levels have moved up and down like a yo-yo. Each time the glaciers advanced, they captured water and sea level dropped as much as 400 feet. As the glaciers melted, the sea level began rising. We are still, by geologic reckoning, in the middle of the Ice Age. In fact, if all the ice on Earth melted tomorrow, sea level would rise at least 180 feet.

FIGURE 2.5
A transgressive barrier island in Kotzebue Sound, Alaska. This narrow island is rapidly migrating in a landward direction into a marsh, gradually filling in the meanders and oxbows of the tidal channels. *Photo by Miles Hayes.*

When each new advance of the ice occurs, the shorelines move seaward, back out across the flat continental shelves. The flooded estuaries are emptied, and the shoreline becomes straight and regular. There is no shoreline to straighten out, so there are no barrier islands.

The last time the sea level was as high as it is today was around 120,000 years ago—at least many believe this to be true. Others argue that the last sea level high was 85,000 years ago, and still others believe it was older than either of these numbers. Geologists in South Africa believe that sea level rose to its present level 120,000 years ago, diminished again, and rose once more, to the current level, 80,000 years ago. To avoid entanglement in the furious and ongoing debate about sea level history, I will, for the sake of simplicity, assume that 120,000 years ago, plus or minus a few thousand, is the correct date of the most recent sea level high.

At that time the shoreline was close to the same level as the present one, a few feet higher in most cases. When the ice again began to spread across the continents, barrier islands were left high and dry, like stranded whales on a beach. And because the two levels of the sea are almost the same, the ancient barrier islands are today located close to the modern barrier islands. All over the world, the former barrier island chain plays a role in determining the shape of lagoons behind barrier islands and occasionally is even incorporated into modern barrier islands (as in Jekyll Island, Georgia).

The sea level began its last major rise, which culminated in the present level, about fifteen thousand years ago. This began the event referred to as the *Holocene sea level rise*. The *Holocene Epoch*, a geologic time unit, began eight thousand years ago and continues to the present. It is the sea level behavior during the last half of the Holocene that has influenced barrier island development. Initially the Holocene sea level rise was very rapid; then about five thousand years ago the sea rose to within a few feet of its present level.

New Zealanders believe that sea level rose to its current level six thousand years ago and stayed put. In Taiwan, the evidence indicates that during the late Holocene, there were two rapid rises (above the present sea level) and falls of the sea level before it settled to its current location. Six thousand five hundred years ago on the Ganges Delta, sea level was ten feet above the present level and the shoreline was 60 to 180 miles further inland.

Sea level has generally risen slowly along the U.S. East Coast during the last four thousand years but may have fallen as much as twenty feet during the same time period along the Brazilian coast (fig. 2.6). The result is a less crooked shoreline along eastern South America, which provides less opportunity for barrier islands to form. The difference in sea level behavior, relative to North America, explains the paucity of coastal plain barrier islands off Brazil and perhaps in the Southern Hemisphere in general.

The reason that the history of sea level change can be different in the Northern and Southern Hemispheres, and even in some extreme cases between adjacent barrier islands, is that sea level change is relative. It involves changes in both the land and the sea. If the land sinks, the sea level rise rate will increase. Perhaps 50 percent of the sea level rise now occurring along the U.S. East Coast (one foot per century) is attributable to sinking of the continental shelf from the weight of the water sitting on it.

Land commonly sinks along the margins of deltas as the buried sediment compacts. Compacting Mississippi mud explains why the sea level is currently rising at a rate of four feet per century off the Delta. Yet sea level is dropping rapidly along portions of the Alaska (Juneau) and Scandinavian coasts as the land rebounds upward from the recent retreat of heavy glaciers.

Earthquakes can cause instantaneous changes in sea level on a regional or a local scale. In Alaska, the 1964 Good Friday Earthquake caused much of the Kenai Peninsula to sink three or more feet along hundreds of miles of shoreline. The Colombian Pacific barrier islands furnish an extreme example of differences in sea level rise rate over very short distances. Because of movements on small local faults, Soldado Island has sunk at least three or four feet in the last few decades, causing rapid shoreline retreat, while the adjacent Santa Barbara Island has remained in place.

The largest changes in sea level are due to the advance and retreat of glaciers, but oceanic currents, when they pile up against or move away from continents, can also cause sea level differences. If the trade winds that push water up on the north coast of Puerto Rico were to reverse suddenly, a sea level drop of several feet would occur. Each time an El Niño happens off Peru and Ecuador, changes in oceanic currents cause temporary two- to four-foot sea level rises at the shoreline. As the sea continues to warm up in response to global warming, sea water in the upper layers of the ocean expands, adding to the global sea level rise.

On the largest scale of all, some of the differences in the history of sea level change from region to region are related to changes in the shape of

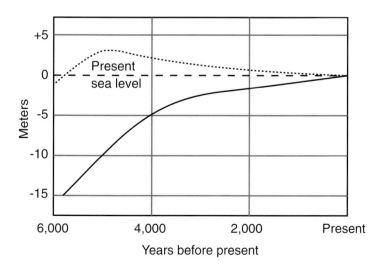

FIGURE 2.6
Comparison of two idealized curves showing the history of the level of the sea during the last six thousand years. Barrier island development and formation are highly dependent on sea level behavior. The lower solid line represents North America, where it is assumed that sea level has gradually risen (perhaps with some local blips) to its present level. The upper dashed curve represents the Brazilian coast and probably most of the Southern Hemisphere, where sea level is assumed to have risen above the present level four or five thousand years ago and then gradually lowered. Detailed sea level history varies widely from place to place and is only poorly understood.

the globe. The almost spherical shape of Earth adjusts a bit as the heavy ice sheets move and as tectonic forces shift the plates about. This change in shape may include the surface of the oceans.

There are strong signs that about one hundred years ago a global sea level rise began anew (fig. 2.7), probably as an early response to global warming. There are also strong indications that barrier islands in a number of places around the world are responding to this sea level rise with increasing shoreline erosion. The barrier islands of the world are narrowing as if they sense and are preparing for the sea level rise.

The Role of the Shoreface

Virtually every grain of sand on a barrier island comes from the island's beaches. Most of these grains at one time or another resided on the shoreface, the submerged lower beach or seaward face of the barrier island. This relatively steep area between the surf zone and the continental shelf acts as a giant sand pump, spewing sand onto the beaches from water depths as great as forty feet. In fact, the shoreface is really the lower beach. The mechanism by which sand comes ashore is not entirely clear, but it is widely assumed that most sand is pushed ashore by waves during times of fair weather or mild wave activity. During storms, sand flow may reverse as beach sand is carried offshore. In the Arctic, sand is largely carried to and from the shoreface by ice.

As islands migrate, the entire shoreface migrates too, and the geologic underpinning of the shoreface may determine the rates of migration and shoreline erosion (fig. 2.8). For instance, if hard rock is exposed (crops out) on the shoreface, both shoreline erosion and island migration rates will be slowed down. If mud crops out, rates may speed up.

FIGURE 2.7
A rim of dead trees lines the contact point between salt marsh and forest along a mainland shoreline of Pamlico Sound, North Carolina. Such dead trees along many coastal plain bays and lagoons around the world are likely a response to rising sea level.

The rate at which such shorefaces can move may be partly controlled by bottom-dwelling organisms rather than wave activity alone. Burrowing and boring clams weaken and even break up rock and compacted mud, greatly facilitating its removal by storm waves. After a storm removes a layer of rock fragments, the clams go to work all over again.

The depth of the base of the shoreface depends on the energy of the waves. Higher waves lead to greater depth of the shoreface base. The depth of the shoreface may also be an important factor in determining island migration rates. The shallower the shoreface base, the smaller the amount of material that needs to be moved with the moving island. Some shorefaces on the rapidly migrating Arctic islands off the North Slope of Alaska are only six or seven feet deep. Morris Island, South Carolina, is rapidly migrating because the effective shoreface, at least during the last few decades, has been less than five feet deep as the shoreline migrates across a hard layer of rock, leaving the island's lighthouse far out to sea.

FIGURE 2.8
The effect of underlying geology on barrier island erosion rates. Here at the village of Rodanthe on the Outer Banks of North Carolina the shoreline is highly irregular. Protruding sections of it are underlain by rock from the Pleistocene epoch. To the north, at the top of the photo, the shoreline curves landward because it is an inlet fill island section and unconsolidated sand underlies the island. *Photo by Mary Edna Fraser.*

The Role of Inlets and Tidal Deltas

Inlets are passages between islands for the waters brought in and out by lunar tides and storm surges, and for the outflow of fresh waters coming from mainland streams and rivers. For a stretch of coast that has several inlets, there is a more or less constant total inlet cross-sectional area, presumed to be just right to handle the in and out flow. The inlets along a coastal segment thus have a constant water volume capacity. If a new inlet opens during a storm, an older inlet nearby usually closes, so the regional inlet flow capability remains the same. The old inlet is no longer needed, and the currents become too weak to maintain it. When the Portuguese constructed a new Ancao Inlet, the old, natural inlet quickly closed.

If the volumes of water are small, inlets will be widely spaced and islands will be long—as is Padre Island, Texas, where the tide range is small and freshwater inflow is negligible. The Georgia Sea Islands are short—the inlets are closely spaced—because the tide range is high, river flow is relatively large, and the total amount of water that must flow in and out is large.

Inlets form in a number of ways, but most often during storms. Storm winds pile up water in lagoons, which then rushes catastrophically across the island and back to the sea once the storm winds reverse direction. The returning water cuts a new channel (fig. 2.9), as happened on Core Banks, North Carolina, in Hurricane Floyd in 1999. It is less common, perhaps, for new inlets to form from the seaward side as storm waves smash through

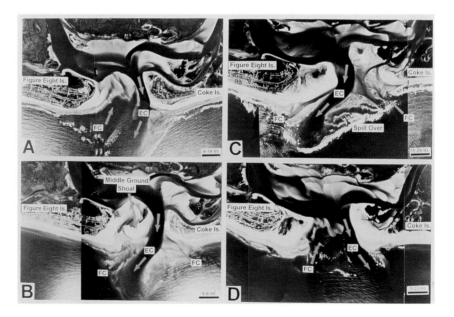

FIGURE 2.9
Changes in the channel configuration of Rich Inlet, North Carolina, between 1986 and 1996. *EC* refers to the ebb channel, and *FC* refers to the flood channels. Note how the position of the shoreline on adjacent figure 2.8 and Coke Island changes as the tidal channels change their location. *Photos assembled by Bill Cleary.*

dunes along narrow island segments, as happened when Hurricane Gilbert (1988) formed more than twenty new inlets across the Mexican Yucatán Peninsula islands. Sometimes, instead of new inlets forming, the old ones widen and deepen during the storm. Oregon Inlet, North Carolina, acted like a cork in a champagne bottle during the great Ash Wednesday storm of 1962, when it widened from half a mile to two miles and deepened from fifteen feet to sixty feet in a matter of hours.

In South Carolina, new inlets have suddenly formed when islands have backed up into tidal channels in the salt marshes behind them during their landward migration. In Taiwan and Colombia, new inlets have opened across barrier islands during quiet high spring tides, without the catalyst of a storm. In the Arctic, new inlets may form as rivers flood in the spring thaw, flow out across the sea ice, and cut their way through an ice-encased barrier island. On some deltas, such as the Niger River in Nigeria and the Rio San Juan in Colombia, inlets are actually river mouths.

Once formed, some inlets migrate while others remain in place. New Topsail Inlet, North Carolina, has migrated at least four miles since it opened in the 1700s, whereas Bogue Inlet, North Carolina, has stayed in one location for centuries.

Inlets typically have large bodies of sand associated with them, called tidal deltas. An *ebb tidal delta*, formed by the outgoing tides, is an arc-shaped body of sand extending seaward of the mouth of the inlet (figs. 2.10 and 2.11). A *flood tidal delta*, formed by the incoming tide, is built out into the lagoon. The sand volume in a flood or ebb tidal delta can be huge, sometimes larger than the total sand volumes of adjacent islands. Along the central South Carolina coast, coastal geologist Miles Hayes estimated that three-quarters of the total

FIGURE 2.10 *(left)*
An idealized diagram of an ebb tidal delta. Note the striking similarity between this diagram and the ebb tidal deltas of some South Carolina Islands in figures 1.2 and 2.11. *Diagram by Miles Hayes.*

FIGURE 2.11
Photo of the ebb tidal delta at Price Inlet, South Carolina. This compares closely with the diagram in figure 2.10. The main ebb channel and the two marginal flood channels can easily be identified in the photo. *Photo by Miles Hayes.*

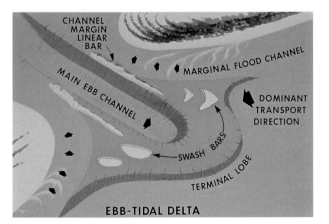

sand volumes in the barrier island systems are submerged in the tidal deltas, both flood and ebb. Sand is carried from one island to the next by a complex process of sand transfer across both tidal deltas. When inlets close, the ebb tidal delta is quickly dispersed by the open-ocean waves, but the flood tidal delta is preserved in the lagoon to eventually become part of the island.

The Role of Vegetation

Because of salt spray, winds, and flooding, island plants occur in distinct zones, parallel to island length. The most-salt-and-wind-tolerant plants are found closest to the beach, and the least-tolerant ones are near the lagoon shoreline. When more vulnerable, less-salt-and-wind-tolerant plants appear along the beach, they are out of place. It is a sure sign of an eroding shoreline (fig. 2.12) that has retreated into what was once the island's interior.

Along the U.S. East Coast, the dominant dune plants control dune shape. Sea oats in the southern United States grow in clumps, resulting in isolated dunes with gaps between them that provide passages for overwash. American beach grass, the northern U.S. dune grass, tends to grow in lines rather than clumps. The result is a continuous frontal dune, with no gaps for overwash, except in the largest storms.

When new sand arrives on the island by storm overwash, plants control what happens next. On Plum Island, Massachusetts, grasses on sand flats between the dunes are killed by sand burial, which leaves the overwash fan vulnerable to wind erosion. On Shackleford Banks, North Carolina, sand flat plants survive burial and grow through the new sand layer, which tends to hold newly arrived overwash sand in place and thus raise the island's elevation.

Island Migration

The incredible process of barrier island migration was first recognized by scientists in the late 1960s and early 1970s. All three of the individuals responsible were students at the time they formulated their ideas—John Fisher at the University of North Carolina, Paul Godfrey at Duke University, and Robert Dolan at Louisiana State University. All three came to the conclusion that islands could migrate, based on their studies of North Carolina's Outer Banks (batik 2.4).

Some, perhaps many, observant people noted that islands could migrate well before scientists formalized the phenomenon. I have been told several times about a "crazy" old man in Nags Head, North Carolina, who, back in the 1950s, claimed that Outer Banks barrier islands migrated. He

FIGURE 2.12

"Out-of-place vegetation" on Hunting Island, South Carolina. Two patches of dark maritime forest, which originally grew where wind and salt spray were reduced, have come to the beach, due to extensive shoreline retreat on this barrier island. The presence of maritime forests adjacent to the beach is a sure global indicator of significant shoreline erosion. *Photo by Miles Hayes.*

would announce this fact at the drop of a hat whenever the subject of island management came up, but he was never taken seriously.

The idea that islands could migrate hit coastal science like a bomb. On a microscale, its impact can be compared with the impact of the plate tectonics theory on the entire field of geology. Coastal science and coastal management would never be the same. In 1972 the National Park Service announced that since erosion of barrier islands was really part of an island migration process, the government was no longer going to attempt to halt shoreline retreat on national seashores. The government would not erect artificial barriers to hold back the sea because there was no longer the worry that the islands would vanish. "Let nature rip" was the cry.

BATIK 2.4

Drum Inlet Nocturne, 1995, 55" x 35"
This view of Core Banks in the Cape
Lookout National Seashore provides an
outstanding example of a migrating
transgressive island. The reddish lobes in
the batik are overwash fans that widen the
island during storms. The fans are now
covered by salt marsh. At Drum Inlet,
well-developed ebb and flood tidal deltas
can be seen. The ebb tidal delta is the
smaller of the two because tidal range is
slight and waves are high.

Artist's note: Drum Inlet became a night
scene in this batik. The mudflats on the
lagoon side and the sediment patterns in
the water are captivating visually. It is
always a challenge to transpose one of my
aerial photographs into a silk tableau. The
island in the foreground is Core Banks.

At the time, the notion was just too radical to find immediate acceptance. Even the normally cool and calm voice of TV news anchor Walter Cronkite dripped with incredulity when he announced the government's new hands-off-the-eroding-shoreline policy. It would be decades before the U.S. Army Corps of Engineers, the agency that manages the nation's beaches, would accept the idea that these strips of sand could save themselves from drowning and move landward as sea level rises.

Islands migrate like the tread of a bulldozer. They retreat on the ocean side and widen on the lagoon side as the island's elevation is raised. They simultaneously move back and up, keeping pace with sea level. The storms that make the shoreline retreat also provide the waves that dump sand in the island's interior and sometimes all the way across into the lagoon. Such washover (for some reason called "overwash" by geologists), simultaneously widens and raises the island (fig. 2.13). But in places where this happens on a regular basis, the islands are usually narrow, less than a couple of hundred yards wide.

FIGURE 2.13
One of the Magdalen Islands in the Gulf of Saint Lawrence illustrates the role of fetch in control of the extent of overwash on a barrier. The fetch for this island is on the order of fifty miles to the east (left) and more than one hundred miles to the west (right). Significant wave activity and overwash occur on both sides of the barrier island, but most overwash (and hence the biggest storm waves) comes from the direction of largest fetch. Note the small and unusual elongate ponds that form perpendicular to the shoreline.
Photo by Miles Hayes.

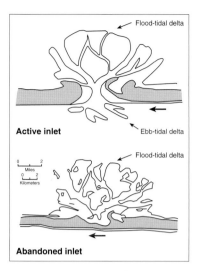

FIGURE 2.14

How barrier islands widen by incorporation of flood tidal deltas. Once an inlet closes, the former tidal delta builds up to sea level by the sand-trapping effect of salt marsh or mangroves and eventually becomes part of the island.

Some islands, such as Tavira Island in Portugal, widen by incorporating the flood tidal deltas left behind when inlets closed (fig. 2.14). Desert barriers, such as South Padre Island, Texas, may widen as sand is blown into the lagoon across broad, gently dipping sand flats. Arctic islands may widen as sand and gravel are pushed across the island by wind-driven sheets of sea ice.

Whatever the mechanism of migration, the mainland shore must also retreat if the islands are to remain islands. That's the reason that ancient barrier islands, especially the 120,000-year-old chain, determine the shape of lagoons. When the retreating mainland shoreline encounters the mass of a former barrier island, the retreat rate suddenly slows down and the mainland shoreline remains more or less in place for a long time. Meanwhile the barrier island in front of it continues to migrate, narrowing the lagoon. Such has happened behind Core Banks, North Carolina, and Padre Island, Texas.

After formation fifteen thousand years ago, at what is now the edge of the continental shelf, barrier islands may have migrated all the way to their present locations. Old inlet and estuarine deposits on the continental shelf, well seaward of today's shoreline, attest to their former presence. During spurts of sea level rise it is likely that the islands disappeared for a while. On very flat continental shelves such as in the Gulf of Mexico, islands were most likely absent during the Holocene sea level rise, because the shoreline moved too rapidly for the islands to keep up. The flatter the continental shelf, the more rapidly the shoreline moves across it in response to a given sea level rise. During times of most rapid sea level rise, shoreline retreat rates on the Gulf of Mexico shelf may have been hundreds of feet per month. It is intriguing to imagine that if Native Americans camped along Gulf beaches eight thousand years ago, they probably had to move their campsite back every day.

THE GLOBAL VIEW

The huge variation in the nature of barrier islands becomes most impressive with the benefit of a global view, even the sketchy one presented here. The islands turn out to be a complex continuum of sizes, shapes, appearances, and origins. A multitude of bodies of sand in a huge range of coastal settings, they have much less in common than one would assume at first glance. But they are all ridges of sand or gravel piled ashore by waves or ice, and all tend to straighten crooked shorelines in nature's never-ending and never quite successful quest for the perfect stable and straight shoreline.

The American Barrier Island Scene
Hot Dogs and Drumsticks

Tonya Haff must have a photographic memory. As my summer student assistant, she was engaged in following up on one of my ingenious ideas—to compare barrier island scenes on postcards of different ages to see how the beaches had fared over past decades. By examining a large antique postcard collection, Tonya sought to document historical changes in beach width (erosion or accretion), seawall construction, and the density of development.

One day, as she sorted the Atlantic City, New Jersey (Absecon Island), postcards, Tonya discovered a card that looked familiar, yet different. On returning to the files, she found that her memory was indeed correct: there was a look-alike card. The two postcards were nearly identical except that one had only a few people on the beach and the other had thousands. The positions of cars and some beach umbrellas were the same on both cards. Clearly one edition had been cleansed to make what was then the world's most famous barrier island seem uncrowded in hopes of thereby attracting even bigger crowds. Tonya's next discovery was that artists had *added* people to the scenes on some cards, always women, always scantily clad (figs. 3.1 and 3.2). All of this was an interesting curiosity, providing a colorful sidebar for my research. Her next discovery, however—that artists sometimes went farther, skillfully widening beaches and improving dunes—dealt a fatal blow to my too-clever study.

Not far behind Atlantic City in abundance of postcards in antiques shops are the eight Long Island, New York, South Shore, barrier islands (batik 3.1), where the great U.S. barrier island chain begins. We will begin our tour of this chain, the greatest in the world, by moving south down the

FIGURE 3.1
A postcard view of a crowded beach at Myrtle Beach, South Carolina, ca. 1920s.

FIGURE 3.2
The same beach view on this postcard has been artfully revised by the removal of much of the original crowd and the addition of a beautiful girl.

Atlantic coast (fig. 3.3) and then onward along the Gulf of Mexico coast. Not included are a few isolated barrier islands in New England—for example, Plum Island in Massachusetts and Orchard Beach in Maine. The 156 islands total 2,100 miles in length and make up 20 percent of the world's barrier islands (as measured by length). Obviously we cannot discuss all of the islands from Montauk Point, Long Island, to the Mexican border, so we will touch down on an occasional island, to illustrate the variety of ways in which islands and their human inhabitants have evolved.

THE MID-ATLANTIC

The New York chain commences with the eroding bluffs at Montauk Point and terminates with Coney Island, the westernmost of the group. Coney Island, however, is no longer an island. Its lagoon and salt marshes were long ago filled in to provide space for an expanding New York City. Nearby Jones Beach, once a tiny sliver of an island, was expanded by the infusion of forty million cubic yards of sand, equivalent to four million dump truck loads, to provide summer beach recreation space for millions of sweltering New Yorkers.

The New Jersey Barrier Islands, commencing just across the Hudson River from Long Island, have had a longer history of interaction with European settlers and their descendants than any other North American islands. Native Americans found the islands useful as a base for harvesting fish and shellfish in the summer but uninhabitable on a year-round basis. These earliest of Americans retreated each year to the forested shelter of the mainland for the winter.

BATIK 3.1
Fire Island, N.Y., 1995, 107″ x 55″
Fire Island is part of the New York barrier
island chain where the Great American
Super-Barrier Chain commences and
extends south beyond the Mexican border.

Artist's note: Fire Island is my art
addressing AIDS as a crisis. The burning
diagonal of the New York barrier island
represents a fiery issue in our country and
holds a hieroglyphic message that has not
been decoded to cure this disease. The
waves to the left beat against the
backbone of my now dead friend. The
calm to the right is love and support of a
spiritual nature.

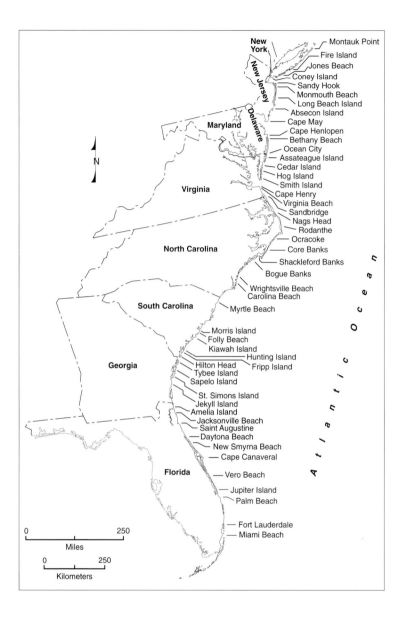

FIGURE 3.3
Map of the U.S. East Coast from Long
Island south.

By the seventeenth century, a few hardy settlers came to the barrier islands, but mainland property still sold for many times the price of island
property. As an example of the disdain that early New Jersey settlers held
for the islands, historian Craig Koedel notes that much of Wildwood Island
was purchased for nine pounds in 1700. The seller used the money to buy
a fine dress for his wife. During the later 1700s, small towns appeared on the
islands, which were now used to harvest timber and graze cattle. The
beachfront began to achieve some value as a place to store and launch boats
to fish or to hunt the nearshore whales. Recreational swimming wasn't

even on the horizon in those days, but it is hard to imagine that people did-n't occasionally stroll on the beach on fine summer days or stand transfixed, as we do today, hypnotized by the tumbling waves.

Then, in 1801, it happened. Postmaster Ellis Hughes of Cape May, New Jersey, placed an ad in the *Philadelphia Daily Aurora*. The ad is credited with beginning the American rush to the barrier islands:

> The subscriber has prepared himself for entertaining company who use sea bathing, and he is accommodated with extensive house room, with fish, oysters, crabs and good liquors. Care will be taken of gentlemen's horses. Carriages may be driven along the margin of the ocean for miles, and the wheels will scarcely make an impression upon the sand. The slope of the shore is so regular that persons may wade a great distance. It is the most delightful spot that the citizens may retire to in the hot season.

Today you'd have a difficult time indeed driving a carriage for even a short distance along the Jersey Shore. Roughly 50 percent of the shoreline of the state is lined with engineering structures, mostly seawalls, giving rise to the term *NewJerseyization*, used to characterize long reaches of wrecked beach, covered by the debris of wave-damaged and deteriorating seawalls and groins. The New Jersey coast has been seawalled far longer than any other in North America, and thus it provides the best place to see the long-term impact of seawalls on a sandy shoreline.

Sea Bright, near the northern tip of the New Jersey coast, is the endpoint of NewJerseyization. During the first half of the nineteenth century it re-mained undeveloped, then in the 1860s a settlement called Nauvoo sprang up, consisting of clusters of fishing shacks and sheds. The first permanent house was built in 1869, and by 1877, when a photograph was taken from a grounded French ocean liner (fig. 3.4), a sparse row of beachfront buildings was present, with no sign of seawalls. The island was low and readily over-washed, however, so by 1886 cottage owners began to put in seawalls. In 1898 a large rubble (stone) seawall was installed for most of the length of the peninsula, yet the beach remained and daily summer trains from New York City continued to disembark passengers within yards of it. This site provid-ed a good illustration of the fact that beach destruction in front of seawalls does not occur instantly. Instead, it takes several decades, which makes it a politically difficult issue to deal with. That is, to change things requires politicians with foresight! The Sea Bright wall gradually grew into a massive seventeen-foot-high cemented stone wall, with no beach in front of it at all until a large beach nourishment project was carried out in the mid-1990s.

An 1877 view from the deck of the grounded French ocean liner *L'Amerique*, offshore at Sea Bright, New Jersey. At the time the island was narrow, low, and without dunes—not a good choice for development. Today Seabright is a walled city with a giant seawall and a nourished beach in front of it. *Photo furnished by George Moss.*

Comparison of nearshore depths on old versus new navigation charts indicates that the shoreface off Sea Bright has grown considerably steeper since the seawall was put in. What this shore steepening suggests is that if the walls were removed tomorrow, the shoreline would retreat extremely rapidly, until the shoreface was once again less steep. And some islands would disappear altogether.

The New Jersey and Sea Bright experience tells us that the choice to armor a barrier island shoreline with seawalls is irreversible. Once they have been in place for a while, they cannot be removed without major land loss.

The New Jersey coast starts out at the north end with Sandy Hook, a still-growing Holocene spit that extends into the mouth of the Hudson River. Sea Bright is the point at which the ten-mile-long spit connects to the mainland. South of the spit is a long headland or mainland shoreline, which, before it was completely seawalled, eroded to provide sand to Sandy Hook. Today, Sandy Hook receives lots of sand from the south, but it comes from eroding artificial beaches rather than from an eroding bluff. Below the headland are nine barrier islands, ending at Cape May. Beach sand transport is from south to north for the northernmost forty-five miles of the coast and from north to south for the remainder of the shoreline. Basically the same sequence of development—a spit, followed by mainland, followed by barrier islands—makes up the shoreline to the south of both Delaware Bay and Chesapeake Bay. A south-to-north reversal of longshore currents occurs as well, just south of both the Delaware and the Chesapeake Bay mouths.

Perhaps it is worth explaining why I usually mention the direction of longshore currents or beach sand transport for each island chain. This critical piece of information about a barrier island's character reveals where the island's sand supply comes from, tells how large it might be, and gives hints as to island origin. Direction also provides the basis for predicting how human activity will affect adjacent beaches and islands.

The Delmarva Peninsula (Delaware, Maryland, Virginia) between Delaware Bay and Chesapeake Bay, with its fourteen (the number can be higher or lower, depending on what one counts as islands) barrier islands, has little development south of Ocean City, Maryland. Perhaps the most spectacular impact of a jetty on the U.S. East Coast is the erosion caused by the Ocean City Inlet jetties. The combined loss of sand trapped on the beach and moved offshore starved Assateague Island and resulted in its rapid migration completely off its 1933 footprint. Today's open-ocean surf zone of Assateague Island is behind its 1933 lagoon shoreline!

Beginning with Metompkin Island and ending at the tip of the Delmarva Peninsula with Smith Island is a ten-island chain, owned by the Nature Conservancy, a private group that purchases land for preservation. This is probably the most pristine barrier island chain in North America south of the Arctic Circle. There is light development on Cedar Island, where the houses are frequently and routinely moved back from the rapidly retreating shoreline.

One hundred years ago Hog Island was a drumstick island (see definition below) with the bulbous end to the south. Today it is a drumstick island with the bulbous end to the north. During the twentieth century, the shoreline rotated as the north end of the island accreted (widened) a full mile. No sign remains of Broadwater, once a thriving Hog Island community (fig. 3.5),

FIGURE 3.5
Gone but not forgotten: the old village of Broadwater, Virginia. The entire village has been lost to erosion. This photo from the 1920s was taken after a rare snowstorm. In the foreground is the backbarrier salt marsh (with dredged channels) and at the top of the photo is the open-ocean shoreline. The steel lighthouse tower in the center of the island was on the beach by 1940 and now resides, as a pile of rubble, more than half a mile from today's shoreline. A few years back, fishers reported catching tombstones from the Broadwater cemetery in their trawler nets. *Photo furnished by Rick Kellam.*

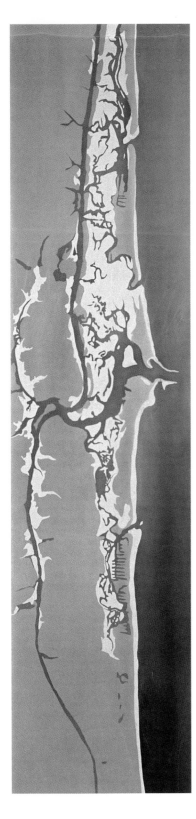

complete with a hotel and post office. Some buildings were moved to the mainland, but most were lost at sea. By the 1950s, fishers reported dragging up Broadwater tombstones in their nets.

Oregon Inlet (fig. 1.3) is the first inlet south of Chesapeake Bay, more than eighty miles away. Within historic time, however, a number of inlets, more than twenty, have existed between the Virginia state line and Oregon Inlet long enough to be named and put on a chart. Some geologists suggest that the larger number of inlets 100 to 250 years ago reflects a stormier time.

THE GEORGIA BIGHT

The Georgia Bight is the long, curved shoreline extending from Cape Hatteras, North Carolina, to Cape Canaveral, Florida. Geologist Miles Hayes calls the bight "the centerpiece for the longest single development of barrier islands in the world . . . Cape Cod, USA to the Yucatan Peninsula, Mexico." The rough state-by-state barrier island count between Hatteras and Canaveral is twenty-one for North Carolina, thirty-three for South Carolina, ten for Georgia, and seven for northern Florida.

Along this 725-mile reach of shoreline, 396 miles, or 55 percent, is lined with true barrier islands. These islands are divided roughly evenly between transgressive and regressive types. The remaining 45 percent of the coastline consists of mainland shoreline segments (Myrtle Beach, South Carolina [batik 3.2] and Kure Beach, North Carolina) and spits extending from the mainland (Garden City, South Carolina, and Carolina Beach, North Carolina).

The Georgia Bight islands are all coastal plain islands. Barrier islands off the mouths of the Santee/Pee Dee (batik 3.3), Savannah, and Altamaha

BATIK 3.2
North of Myrtle Beach, S.C., 1996, 60" x 15"
The large jettied inlet is Little River Inlet at the boundary between North and South Carolina. The first island in South Carolina is undeveloped Waites Island. The peninsula extending north from the mainland south of Waites is occupied by the town of Cherry Grove, the beginning of the Grand Strand of South Carolina; one of North America's most heavily developed shoreline stretches. The Intracoastal Waterway is the long, continuous channel behind the shoreline.

Artist's note: When I was a child, the white sandy beaches between Windy Hill and South Myrtle Beach were my family's favorite summer vacation destination. They were family beaches, with cottages and very few high-rise condominiums. My cousin commissioned this batik, which holds fond memories.

Rivers protrude slightly seaward but are not true delta islands. The Georgia Bight islands receive almost no sand from their rivers at the present time because the river sand load is dumped at the heads of the estuaries, miles from the beach. The Georgia islands are composite islands, combined Pleistocene and Holocene sand bodies (chapter 11).

In the summer of 1999 an event took place at the north end of the Georgia Bight that had huge implications for barrier island management in North America and perhaps around the world. The erosion-threatened 3,500-ton Cape Hatteras Lighthouse on Hatteras Island (fig. 3.6) was moved landward out of harm's way 1,600 feet from the beach and positioned at the same distance from the sea as when the lighthouse was constructed in the 1880s. The saga began in 1979, when the lighthouse nearly fell victim to a storm and was saved only when the National Park Service, in the middle of the storm, tore up a parking lot, and dumped it into the sea in front of the lighthouse. If ever there was quick bureaucratic thinking, that was it.

An intense twenty-year debate ensued. Options were to move it, seawall it, nourish the beach in front of it, or let it fall into the sea. Local commercial interests opposed the move, probably because they did not want to draw international attention to the erosion problem on the Outer Banks barrier islands. The opponents found experts to testify that the lighthouse could be saved in situ and that the structure was too fragile to move. Finally a joint panel of the National Academy of Sciences and the National Academy of Engineering saved the day and saved the lighthouse with their declaration that not only was it feasible to move it safely but it had to be moved if it was to be saved.

BATIK 3.3
North and South Santee, S.C., 1989, 60″ x 180″
Here two branches (distributaries) of the Santee River arrive at Cape Romain, South Carolina. This cape is made of several small barrier islands, including rapidly changing South Island and Cedar Island, in the foreground.

Artist's note: A flight over the Santee River north of Charleston with my father piloting the Ercoupe made me aware that inlets are as individual as snowflakes. I decided to use my art to promote the awareness of environmental beauty and change on the planet as seen from the air. This silk looks like an ancient kimono and depicts the aerial environs of a textile mill.

This aerial was taken June 22, the sixth day after the move began. The inset, above right, was taken June 27, approaching midway on the move corridor, placing the lighthouse out of harm's way at 1,000 feet, eleven days into the move.

Thanks to relocation –
At the first dawn of the twenty-first century the Cape Hatteras Light Station
will look as it did in the last rays of sunset of the nineteenth century.

FIGURE 3.6
The 3,500-ton Cape Hatteras Lighthouse on the move in the summer of 1999. *Photo courtesy of the U.S. National Park Service.*

From north to south the Georgia Bight has five capes: Cape Hatteras, Cape Lookout, and Cape Fear (batik 3.4) in North Carolina, Cape Romain in South Carolina, and Cape Canaveral in Florida. Associated with each cape is a massive shoal of sand extending straight offshore for distances as great as ten miles (fig. 3.7). The shoals, famous for the large numbers of ships that have run aground on them, are long-submerged ridges along which sand moves from the beaches to the continental shelf, never to return.

At least two of the capes, Lookout and Canaveral, are in about the same location as Ice Age capes. Merritt Island, behind Canaveral and site of the space shuttle launchings, is such a Pleistocene cape. Many hypotheses have floated by to explain the origin of the capes, but none seems to be the definitive answer. They probably have multiple origins, and surely they have

BATIK 3.4

Cape Fear, N.C., 1996, 48″ x 35″
An oblique view of Cape Fear, North
Carolina, one of five major capes on the
U.S. East Coast, from north to south,
Capes Hatteras, Lookout, Fear, Romain,
and Canaveral. The dark green areas in
the batik are the high-forested beach
ridges, or lines of sand dunes.

Artist's note: Cape Fear is the river that
runs through my hometown of
Fayetteville, North Carolina, and flows
into the Atlantic Ocean. The serene
barrier island is Bald Head Island at the
mouth of the river. Taken from one of my
aerial photographs, the batik exemplifies
an excursion flight with my brother,
Claude Burkhead III, as pilot of the
classic Ercoupe.

been strongly affected by the ups and downs of sea levels during the ice
ages, including the sea level rise of the last fifteen thousand years.

Perhaps the capes began as a large, resistant rock outcrop on the continental shelf that held the shoreline in place during a sea level rise, while
adjacent shorelines moved back. This differential retreat is happening today
at Rodanthe, North Carolina, on the Outer Banks, where Wimble Shoals
has at least momentarily retarded shoreline erosion and formed a mini cape.

FIGURE 3.7
A view of the tip of Cape Lookout on the Outer Banks of North Carolina. The position of the Cape Lookout shoals, the large body of sand that extends offshore from the cape, is marked by the breaking waves. *Photo by Mary Edna Fraser.*

Another hypothesis is that the cape and the shoals are old river deltas where sand has been left behind on the continental shelf. One problem with this idea is that no river of consequence exists near Cape Canaveral. Others believe that particular wave conditions produced by waves arriving in the surf zone, at a large angle to the shoreline, encourage sand accumulation that leads eventually to a cape.

Reflecting regional differences in tide range and wave height, the islands at the north and south ends of the Georgia Bight average twenty miles in length and those in the central part of it are much shorter, usually less than ten miles long. Most of the thirty-three South Carolina islands are less than twelve miles long. A few islands, including the islands of Cape Romain (fig. 11.2) and Masonboro Island, North Carolina, are actively migrating toward the mainland. Many others are in the first stage of island migration, thinning down by erosion.

64

THE AMERICAN BARRIER ISLAND SCENE

Miles Hayes recognized that tide range and the height of the waves work in tandem to produce two prominent kinds of barrier islands, which he rather informally dubbed *hot dogs* (the foot-long variety) and *drumsticks*, based on their shapes as seen from the air. These easy-to-visualize terms can be applied as a simple classification scheme around the world. In the Georgia Bight, hot dog islands are best represented by those on the Outer Banks, where the islands are relatively low and long and are more or less the same width everywhere along their length. They are long because the tide range is small, which leads to widely spaced inlets.

Drumstick islands, such as Folly, Kiawah (fig. 3.8), and Bull Islands, South Carolina, are shaped like the leg of a chicken. One end, always the updrift extremity, is bulbous, and the downdrift end of the island is narrow. In order for a drumstick to form, a sizable ebb tidal delta is usually required, which is one reason that the Outer Banks island chain of North Carolina does not include drumstick islands. Wave heights are too high and tide ranges are too low to favor the formation of large ebb tidal deltas.

Drumstick islands form in South Carolina as incoming ocean waves, especially from the east or the northeast, and bend or refract as they feel the bottom in front of the ebb tidal delta (fig. 3.9). The waves bend around the delta and lead to a reversal of the normally north-to-south longshore currents, over a short shoreline distance near the inlet at the north end of the islands. The reversed current causes a sand pile-up just south of each inlet, forming the fat end of the drumstick. The narrow end of the drum-

FIGURE 3.8
An infrared view of Kiawah Island, S.C. The red-colored forested ridges are former beach ridges created when the island was going through a regressive phase and building seaward. At present, Kiawah, like most East Coast barrier islands, has a retreating shoreline. *Photo by Dennis Hubbard.*

stick forms as a result of spit growth to the south (fig. 3.10) by the sand that escaped entrapment at the fat end.

Events associated with tidal deltas are responsible for more than just the drumstick shape of islands. Main channels wander about within the confines of the inlet, and the position of the main channel often determines patterns of erosion and accretion on adjacent islands. In one particularly spectacular case, the north end of Edisto Island eroded back more than a half mile, entirely because of the position of the main channel, which directed sand seaward and away from Edisto, the downdrift island. As a result, the seaside village of Eddingsville fell into the sea in the 1890s.

The saga of Morris Island, just south of the entrance to the Charleston, South Carolina, harbor (batiks 3.5 and 3.6) is a famous example of the extremely dynamic nature of barrier islands and the role of humans in their evolution. Morris Island was the site of the ferocious Civil War battle for Fort Wagner, the final desperate battle depicted in the movie *Glory*. A single concentrated cannon volley from Union Army positions on Folly Beach virtually erased the original Morris Island Lighthouse, and in 1874 a new one was built, 2,700 feet back from the shoreline. In the 1890s, jetties were built

FIGURE 3.9
The origin of drumstick barrier islands. The incoming waves bend around the tidal delta and cause the longshore current to reverse at the end of the downdrift island. The reversed current causes one end of the island to build up to become the meaty end of the drumstick. The main (ebb) channel occupies the center of the inlet. Arrows mark the flow of the flood tides in this greatly simplified diagram. *Diagram by Miles Hayes.*

FIGURE 3.10 *(opposite)*
An infrared photo of Captain Sam's Inlet at the south end of Kiawah Island. The successive recurved spits reflect a history of spit growth to the south. Periodically, a storm will create a new inlet to the north, where the island is narrowest, and the whole north-to-south spit growth process begins anew. *Photo by Miles Hayes.*

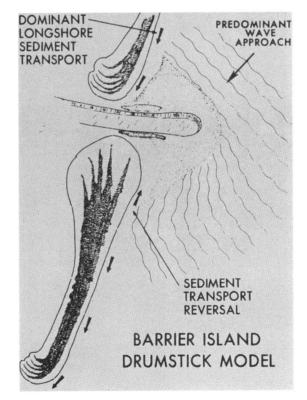

THE AMERICAN BARRIER ISLAND SCENE

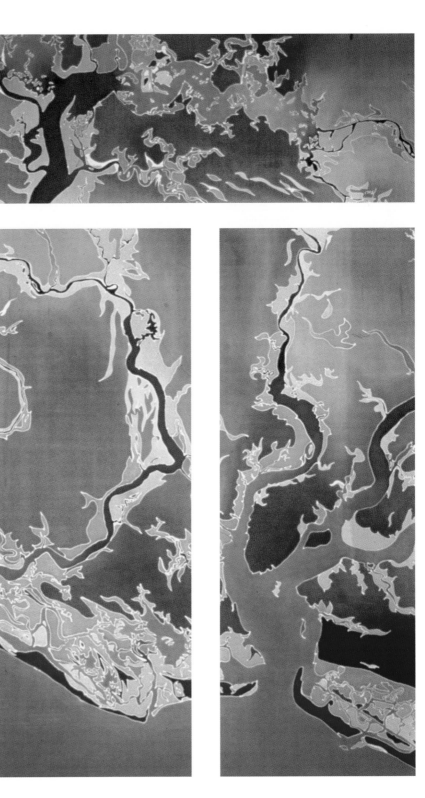

BATIK 3.5

Charleston Red, S.C., 1988, 108" x 72"
The entrance to the Charleston harbor with Kiawah, Folly, and Morris Islands to the south (left) and Sullivan's Island and Isle of Palms to the north. This entire stretch of coast built out rapidly for thirty years, when the huge ebb tidal delta broke up because of the construction of the jetties at the Charleston harbor entrance in the 1890s. Once the delta sand was dispersed, very rapid erosion ensued on Morris and Folly Beach Islands.

Artist's note: This batik is the first time I used navigational charts as sources of imagery. I cut a chart of Charleston into pieces and played with different arrangements. The resulting batik is mobile and here places the North Edisto River and Wadmalaw Island above Charleston instead of in their actual locations southwest of the city. At one time the silks graced the U.S. ambassador's London residence.

to help revive the port of Charleston and the economy of the South Carolina lowlands, which were still recovering from the war. With the jetties in place, tidal currents could no longer maintain the huge ebb tidal delta, which immediately began to break up. Adjacent islands to both the south and the north were furnished with large slugs of sand. But it was a one-time blessing. In the 1930s, after more than thirty years of island accretion, the tidal delta was gone and the situation suddenly reversed. Islands to the south, the downdrift direction, began to erode rapidly, Morris Island most of all.

Unlike the Cape Hatteras Lighthouse in North Carolina, which has a shallow 6-foot foundation, the Morris Island Lighthouse foundation extends 35 feet below sea level. By 1940, the lighthouse was on the beach, and today it stands slightly tilted more than 1,500 feet out at sea (fig. 3.11). But Morris Island has not disappeared. The island migrated landward, out from under the lighthouse.

Folly Beach, the next island to the south (batik 3.7), bears the name given it by mariners, who looked for the island's tree-crested dune ridge, a volley or folly of trees, as a navigation guide to the entrance to Charleston. Probably a lot of East Coast islands bore the temporary name of Folly Beach. The navigation channel into Charleston, the natural main channel of the Charleston ebb tidal delta, used to wend its way in front of Folly Beach, giving rise to another name; Coffin Land. Captains of sailing ships put their sick passengers ashore on Folly, in order not to risk quarantine of crew and cargo upon arrival at the Charleston docks. Returning to sea, the sick were either abandoned, picked up again, or, if need be, buried. During the Civil War twenty thousand Union troops lived on the island as they prepared for the assault on Fort Wagner. To avoid the hazard of idle hands and minds, the troops were kept busy digging drainage ditches (to get rid of malarial mosquitoes) and constructing roads and causeways (some of which are still visible today).

BATIK 3.6
Charleston Runner, S.C., 1996, 15" x 60"
A broad view of the South Carolina coast centered on Charleston Harbor and extending from Edisto Beach to Bull Island on the south limb of Cape Romain. Clearly illustrated here is the mixing of river drainage with the complex combined river and tidal drainage channels in the marshes of the outermost coastal plain.

Artist's note: A small-scale framed work, *Charleston Runner* belongs to the Gibbes Museum of Art. The coast is quietly rendered to look like trees in a misty setting the color of Spanish moss. A nautical chart was used for reference.

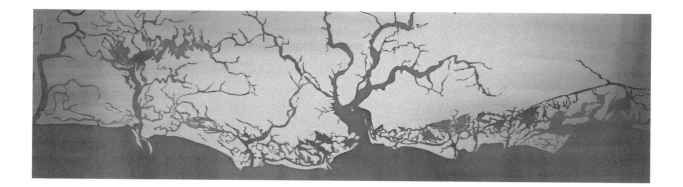

Erosion rates from the 1930s on were so rapid that Folly Beach did not become a habitat for the rich like so many of its neighboring islands. Movable or disposable shacks lined the beachfront, and for some time it was referred to as the hippie beach, a place for upstanding people to avoid. Hurricane Hugo in 1989 changed that. Severely damaged beachfront houses were replaced with upscale million-dollar homes, and Coffin Land began its trek to respectability.

As for the remainder of the Georgia Bight, we will discuss the Georgia islands in chapter 11 and will include the northern Florida islands in a discussion of east Florida.

EAST FLORIDA

The barrier islands of east Florida are a different breed. Some argue that they shouldn't be considered islands at all. Only the northernmost two, near the Georgia border, Amelia and Little Talbot Islands, are unequivocally classical barrier islands with natural inlets and a healthy saltwater lagoon behind them. Much of the remaining coast has an artificial aspect.

One hundred years ago a 150-mile-long barrier island extended, on either side of Cape Canaveral, from Ponce de Leon Inlet near Daytona Beach to Jupiter Inlet near Palm Beach. It was the longest barrier island in the world.

Today, four artificial inlets cut across it. A hundred years ago, there were only eleven natural inlets along the 365-mile east Florida coast between the Georgia state line and Miami Beach. Today there are nineteen, including seven of the original eleven. All inlets except Matanzas and Nassau Sound are stabilized by jetties. Of the twelve inlets south of Cape Canaveral only three (Jupiter, Hillsboro, and Miami River) are natural. The other ten were dredged, or in many cases blasted through rock, mostly in the 1920s.

The saga of Palm Beach, now the "wealthiest barrier island in America," is typical. The "lagoon" behind Palm Beach was a freshwater body called Lake Worth. In 1917 Lake Worth Inlet was opened by mule and drag pan, and South Lake Worth Inlet was opened in 1927 when, for the first time, Palm Beach became an island.

The east Florida islands are backed by lagoons that have numerous origins. Behind some islands are freshwater marshes. At one time the marshes could be waded to get to the islands, albeit with some difficulty—especially since you might need to dodge alligators. Others are open-water lagoons with river or lake names from their former existence. Examples include the Halifax River behind Daytona Beach, the Indian River behind Melbourne

BATIK 3.7
Bird Key, S.C., 1997, 38″ x 65″
Bird Key is a sand shoal or bar on the ebb tidal delta of Stono Inlet between Folly Beach and Kiawah Island, South Carolina. Once the largest brown pelican rookery in the southeastern United States, the shoal disappeared, probably because the sand transported to the inlet from the nourished beach on Folly Beach forced the main channel to move and to erode the island.

Artist's note: Flying and photographing for two decades, I have witnessed the closing of inlets and the disappearance of islands. Bird Key was a brown pelican rookery, but because of dredging, it no longer exists. Luckily a few relocated pelicans took up residence near James Island Creek and entertain us with their dive-bomb fishing by my dock.

Beach, Lake Worth behind Palm Beach, and Lake Mable, which became the Port Everglades turning basin near Fort Lauderdale. Today the Intracoastal Waterway is behind the entire coast and sometimes constitutes the only reason for the designation *island*.

The southeast Florida coast is a rocky coast covered with a thin layer of sand. Well, most people wouldn't call it a rocky coast, but Florida coastal geologist Charles Finkl does. Underlying many of the barrier islands and most of the beaches is limestone—hard Pleistocene rock (Anastasia Formation) originally deposited as barrier islands or coral reefs, believed to have formed during the 120,000-year sea level high. So the present barrier island chain coincides exactly with the ancient barrier island chain. For much of its length here, the Intracoastal Waterway behind the islands was blasted rather than dredged. According to Finkl, the typical thickness of sand over the rock is a mere three to six feet.

Rock-cored islands cannot migrate. If sea level continues to rise, the current island chain will be "overstepped" and a new island chain will form on what is now the mainland. Hutchinson Island, Singer Island, Hillsboro Beach, Fort Lauderdale Beach, Dania Beach, and Miami Beach are all examples of "rock-cored" islands. On Palm Beach a large ridge of coral reef limestone forms the spine of the island. A road cut on the island through this limestone is said to be the deepest road cut in all of Florida!

The backbarrier lagoons in Florida, as those all over the world, are among the world's most productive ecosystems. Largely responsible for this productivity are the two major types of lagoon plant communities: *mangroves* and *salt marshes.* Cape Canaveral, Florida, is the approximate boundary between the two along the east coast of North America. To the south, mangroves predominate behind the barrier islands. To the north, salt marshes are dominant.

Salt marshes are a thick mat of salt-tolerant grasses that exist on sand and mudflats subject to periodic covering by salt water (fig. 3.12). Mangroves are trees that occur as strips or patches on flat intertidal surfaces (fig. 3.13). Both play a very important role in barrier island evolution by stabilizing lagoon shorelines, preventing their erosion. At the same time, mangroves and salt marshes are land builders. They dampen waves and currents, which in turn causes sediment to accumulate, raising the elevation of the intertidal surface to the point that it eventually adds to the land area of the barrier island, an important part of the island migration process.

Both mangroves and salt marshes support the lagoon food chain, including fish, birds, and numerous invertebrates. They act as nursery areas in the life cycles of many marine organisms and provide vast amounts of nutrients that drain to the surrounding waters with each ebbing tide.

Mangrove communities are referred to as *mangrove forests* or *mangrove swamps*. Around the globe there are perhaps fifty to sixty species of mangroves, divided into four main genera: red mangroves (*Rhizophora*), black mangroves (*Avicennia*), white mangroves (*Laguncularia*), and button mangroves (*Conocarpus*). In the Eastern Hemisphere there are five times as many species as in the Western Hemisphere. For example, only three species are found in North America, but twenty-seven have been identified in Australia. Although mangroves have long been a component of the

FIGURE 3.12 *(above)*
A vast expanse of salt marsh along the margins of Pamlico Sound, N.C. In the foreground is the edge of a drainage ditch along a highway. The marsh consists of a single plant species.

FIGURE 3.13
Mangrove forest along the inner edge of a mudflat on the Pacific Coast of Colombia. The mangrove trees are advancing seaward out on the flat. Mangrove forests are the warm-water equivalent of salt marshes, but unlike salt marshes, several species of mangroves may occupy the same intertidal zone. *Photo by Bill Neal.*

west Florida barrier islands, the extensive mangroves lining east Florida's lagoon shorelines have only recently been introduced, because of the creation of new inlets that allow salt water to penetrate the lagoons.

Salt marsh consists of one species of plant in North America, the smooth cordgrass (*Spartina alterniflora*). *Spartina* has no plant competition, as it occupies the zone of daily saltwater flooding by the tides. One variant of *Spartina* that resides next to channels can be as tall as nine feet, but more often it is one to three feet high. Salt-marsh hay was once valued as livestock feed, and it (unfortunately) remains a favorite of wild horses that roam barrier islands.

THE U.S. GULF OF MEXICO

The Gulf of Mexico (fig. 3.14) is quite different from the East Coast. The biggest waves come from the south, but the high-latitude nor'easters that generate the big Atlantic storms have little impact. As a result of the generally lower waves, hurricanes are of huge importance in the Gulf and are the principal agent of change in barrier islands.

The Gulf of Mexico shoreline is bounded by a broad, gentle continental shelf, which saps the energy from any big open-ocean waves. Both tides and wave heights are, on average, small along the entire Gulf Coast. The rough barrier island count is as follows: the west coast of peninsular Flori-

FIGURE 3.14

Map of the U.S. Gulf of Mexico shoreline.

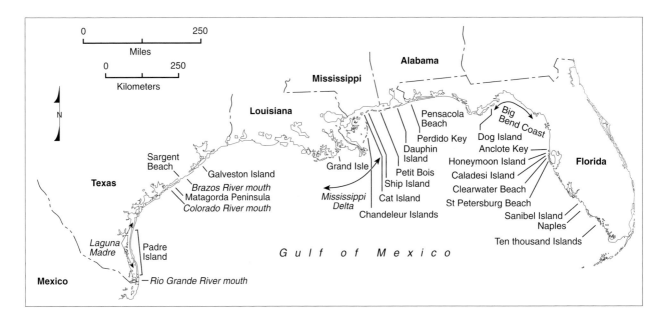

da has thirty barrier islands; the Florida panhandle, Alabama, and the Mississippi coasts (fig. 3.15) are lined with fourteen; the Mississippi Delta of Louisiana (discussed in chapter 5) is bordered by dozens of ephemeral and rapidly changing islands; and Texas has six.

Some of America's most prosperous and densely developed barrier islands lie along the Florida Peninsula coast and include Marco Island, Naples, Sanibel Island, Sarasota Beach, Bradenton Beach, and St. Petersburg Beach. The Everglades and Ten Thousand Islands are found at the south end of this 180-mile-long chain. At the north end is the so-called zero energy coast of Florida's big bend. Here the coast is lined with salt marsh and has no barrier islands. All the requirements for barrier islands are there (sea level rise, flat surface, and sand supply), with the exception of waves, which are not high enough to move much sand. Since the biggest waves tend to come from the open water to the south, the main direction of sand transportation on the Florida Gulf Peninsula is south to north.

Caladesi Island is one of the few remaining undeveloped islands along Florida's peninsular west coast. It is south of Honeymoon Island (to which it was once attached) and north of Clearwater Beach Island, to which it has recently become attached. The island's history illustrates the dependence of west Florida island evolution on hurricanes. In the case of Caladesi Island, the hurricanes of 1921 and 1985 (Elena) had the most impact.

In the 1880s Caladesi Island was called Hog Island. Then the hurricane of 1921 cut the island in half, and the new islands were designated North Hog and South Hog. This must be one of the most common barrier island names in America—at least it was in the old days, before a romantic name became a necessity for real estate sales. Who would buy a million-dollar condominium on North Hog Island? When developers began eyeing North Hog for its potential for beachfront high-rises, the name was changed to Honeymoon Island, and South Hog Island became Caladesi Island (a name whose origin is lost in obscurity).

Caladesi Island sits atop hard limestone rock that is widely exposed offshore at the surface of the continental shelf. The thickness of the sandy island atop the limestone varies between five and twenty feet. Coastal geologist Richard Davis, an expert on the west Florida islands, believes Caladesi Island was in place about 2,500 years ago.

The behavior of Dunedin Inlet provides a great example of the role of the tidal prism in island evolution. The tidal prism is the volume of water that comes in and out of the inlet with the tides. The construction of causeways from the mainland to nearby islands reduced the tidal prism because they effectively reduced the lagoon area. By the 1940s, the reduced flow through the inlet caused the five-hundred-yard-wide, fifteen-foot-deep inlet to narrow to one hundred feet, and shoal (shallow) to ten feet.

At the same time, the inlet migrated to the north. In the perpetual battle between tidal flushing and the infilling of the inlet with beach sand transported by longshore currents, the balance changed in favor of beach sand. Dunedin Inlet began to migrate because the ebb flow of water out to sea had been weakened so that the longshore transport from the south began to deposit sand in the inlet, forcing it to move north.

Then along came Hurricane Elena in 1985. The whole ebb tidal delta at Dunedin Inlet was removed at a single stroke. After the storm, sand that would previously have been transported across the tidal delta and passed to the next island now was pushed into the throat of Dunedin Inlet. The inlet soon closed completely (fig. 3.16), and Clearwater Beach and Caladesi Island became a single barrier island that retained two names.

The last island to the north along the Florida Peninsula is Anclote Key, and the first true barrier island along the Florida panhandle is Dog Island

FIGURE 3.16
Aerial view to the north, of the former
Dunedin Inlet on Caladesi Island. The inlet
closed because causeways across the
lagoon reduced the volume of water that
flowed into and out of the lagoon through
the inlet. *Photo by Richard Davis.*

(batik 3.8). The panhandle coast has eight islands (fig. 3.17), two capes
(Cape San Blas [fig. 3.18] and Cape St. George), three spits, and one mainland reach. There are six islands and several spits fronting the Mississip-
pi/Alabama coast.

Dauphin Island is Alabama's only barrier island completely within the
state's boundaries. Alabama used to have two such islands, but the second
one, Petit Bois (fig. 3.19) is now entirely within the state of Mississippi.
Between the 1860s and the present, Petit Bois Island crossed the state line
and changed its political affiliation by lateral migration.

Fifteen miles long, low and narrow (mostly less than two hundred or
three hundred yards wide), Dauphin Island is a composite island. It probably formed in place as a remnant of the 120,000-year-old ice age barrier
island that was surrounded by water as the sea level rose. The ice age remnant now forms the relatively wide and forested east end of the island, and
its erosion furnishes sand for the island to extend as a spit to the west.
Today the westward growth of the island continues at more than one hundred feet per year.

Dauphin Island has a long and spectacular hurricane history. In the
early part of the twentieth century a hurricane carved a five-mile-wide
shallow inlet, splitting the island in two. In 1948 a hurricane created another inlet. In both storms almost the entire island was overwashed.

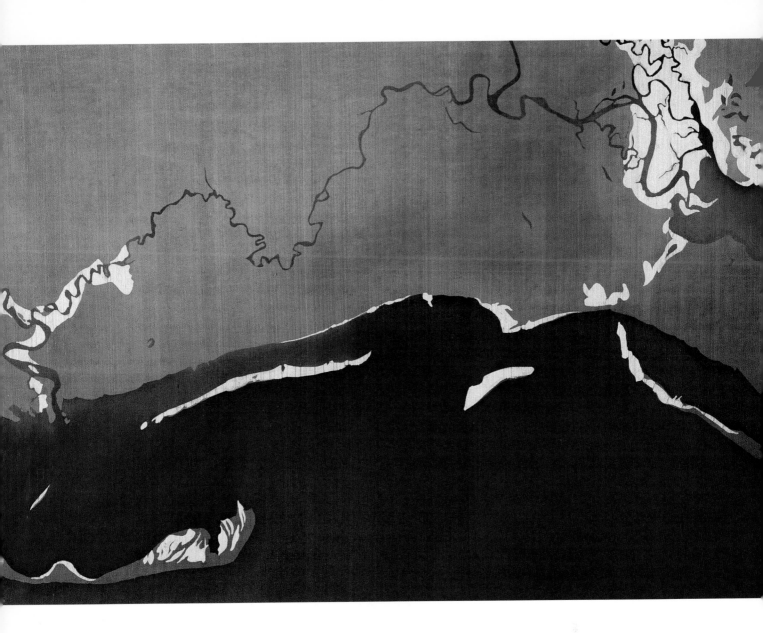

BATIK 3.8

Florida's Dog Island, 2000, 49" x 66"

Dog Island is believed by some to have formed as a sandbar that emerged during a storm. It is the easternmost island of the Florida Panhandle. To the south and east, wave energy is extremely low, a result of the very wide, shallow continental shelf that dampens the waves as they come ashore. The small waves can't build barrier islands, hence the gap in islands between Anclote Key on the Florida Peninsula and Dog Island on the panhandle.

Artist's note: Florida's Dog Island is on the Gulf Coast of Florida and is a commission from a client who wanted the batik dyed to match the colors of an Audubon print of a heron. The art is based on a navigational chart and highlights Florida's Crooked River.

FIGURE 3.17
Santa Rosa Island on the Florida Panhandle immediately after Hurricane Opal in 1995. This island has been narrowing in recent decades as a result of shoreline retreat on both the lagoon and the ocean sides, presumably in response to rising sea level. Barring attempts by humans to hold the island in place, it eventually will begin true migration. This will happen when it is sufficiently narrow and low for overwash to flow across it on a broad front at frequent intervals. *Photo by Greg Stone.*

FIGURE 3.18
Saint Vincent Island on the Florida Panhandle, showing the beach ridges formed when the island was building seaward. *Photo by Greg Stone.*

FIGURE 3.19
Overwash fans crossing Petit Bois Island, Mississippi. Vegetation is just beginning to reappear on the three distinct lobes of sand that cross the island. *Photo by Michael Hobbs, Gulf Islands National Seashore, U.S. National Park Service.*

FIGURE 3.20
Dauphin Island, Alabama, after Hurricane Frederic, 1979. The entire island was overwashed, and damage to buildings was extreme, from the forces of both the sea and the wind.

Eventually the inlets closed naturally. In 1979 Hurricane Frederic struck Dauphin Island (fig. 3.20), did immense structural damage to buildings on the now rapidly developing island, and destroyed the bridge leading to the island. Dismayed by the damage, the State of Alabama, for a brief spell, considered abandoning the island and not rebuilding the bridge. The reflective moment passed, however, the forces of capitalism prevailed, and buildings were reconstructed, while a new $50 million causeway was built. Development had reached a new peak by the time Hur-

ricane Danny struck in 1997, destroying buildings by overwash and dropping thirty-six inches of rain on Dauphin Island, twenty-six inches of which fell in seven hours. Damage was once again immense, and once again rebuilding took place.

Mississippi's islands are narrow, low, and ephemeral (fig. 3.21). Dog Island (another popular name for barrier islands), near the western end of Horn Island, off the Mississippi coast, no longer exists. In the 1930s, however, the island hummed like a beehive, with a thriving dance pavilion, casino, restaurant, cabanas, piers, and docks. In the best tradition of creative entrepreneurship, developers changed the name Dog Island to Isle of Caprice. Alas, erosion and hurricanes took it all away, leaving only an artesian well flowing through a pipe, protruding from the ocean, where for years fishers pulled up to grab a drink of cool, fresh water at a point where no land was visible on the horizon.

FIGURE 3.21
A grayscale infrared NASA aerial photo of nine-mile-long Ship Island, Mississippi. Light areas in the photo are lightly vegetated sand dunes and beaches; gray regions are low, often marshy, well-vegetated dune flats with small dune ridges. The long narrow black areas are cat's-eye ponds, elongated patches of fresh or salt water between small beach ridges. Such ponds are a common occurrence on the Gulf of Mexico barrier islands of Florida, Alabama, and Mississippi. The bottom of the photo is the open-ocean side. On the lagoon side of the island a complex pattern of submerged sandbars is clearly visible. *NASA photo furnished by Michael Hobbs.*

TEXAS

The Texas barrier islands are both coastal plain and delta islands that obtain their sand from the Rio Grande, Brazos, and Colorado Rivers. Unlike the rivers of the U.S. east coast, those in Texas supply sand directly to the islands, or at least they used to. The Texas coast has six barrier islands of the hot dog variety and two barrier spits, the Bolivar and the Matagorda Peninsulas. From northeast to southwest, the islands are Galveston, Follets, Matagorda, San Jose, Mustang, Padre, and Boca Chica. Two of the barrier islands are particularly famous; Padre because it is now the longest barrier island in the world and Galveston (fig. 3.22) because it was the site of North America's greatest historical natural disaster ever. The 1900 hurricane, chronicled in the bestseller *Isaac's Storm*, killed at least six thousand Texans (counting fatalities on the mainland, there may have been as many as ten thousand deaths).

The famous 1900 hurricane has not halted Texas island development. Those were the days before good hurricane predictions could be made, but the American Weather Service ignored a hurricane alarm from Cuban weather experts. The Cubans used such "crude" methods as observations of cloud patterns instead of modern science to warn their people of impending danger. In spite of the magnitude of the disaster, the Galveston survivors stayed on, but they took precautions. A massive seawall was built, and of greater importance, the elevation of their city was raised with a large infusion of sand. What had once been street-level floors on multi-story buildings became basements (fig. 3.23).

Strange things reside on barrier islands. One of these is the "Biosafety Level 4 Laboratory" on Galveston Island, where many of the world's most dangerous viruses will be studied, including Ebola, West Nile, Rift Valley fever, and yellow fever. In case of a big hurricane, the lab will be instantly decommissioned and all viruses and laboratory animals destroyed, or so it is hoped. Another inexplicable object on a barrier island is the nuclear power plant on Hutchinson Island, off the east coast of Florida.

FIGURE 3.22
Aerial view of Galveston Island showing subtle beach ridges trending parallel to the shoreline. The beach ridges mark former shoreline positions left behind as the island widened. The concept of a regressive island was developed during geologic studies on this island. *Photo by Robert Morton.*

According to Bob Morton, who has studied Texas islands over a lifetime, when the Holocene sea level rose to its present location, sand supply was large and the three rivers constructed large deltas protruding into the Gulf of Mexico. Later, when the climate became drier, less sand came down the rivers, and the barrier islands immediately in front of the river mouths changed from a building-out mode (regressive islands) to one of migrating landward (transgressive islands). The sand from the eroding, sand-starved islands at the river mouths moved laterally via longshore currents to build up the islands between the river mouths.

It is a strange situation. The shoreline in front of the river mouths, where the sand supply might be expected to be high, is eroding, and islands away from the river mouth are building out. For example, the highly erosive Matagorda Peninsula fronts the mouth of the Colorado River, while Matagorda Island, to the south, away from the river, has a long history of building seaward.

FIGURE 3.23

Galveston, Texas, in the process of being raised after the catastrophic hurricane of 1900 that killed six thousand people. Most of the deaths were attributed to the storm surge. Raising barrier island elevations is a good way to reduce this hazard in future storms (though a better way might be to move off the island). *Photo courtesy of the Galveston District of the U.S. Army Corps of Engineers.*

Today the heavy hand of humans on the Texas coast is everywhere. Most of the shorelines of Texas are eroding, because of the usual problems of channel dredging and jetty construction at inlets. Dams and tens of thousands of ranch ponds for cattle in the river drainage areas have further reduced the already small sediment supply from the three rivers. Changes to the mouth of the Brazos River made to facilitate harbor improvement have completely cut off the sand supply to the south. As a direct result, Sargent Beach, about forty miles to the southwest, was suffering erosion rates of sixty to one hundred feet per year (fig. 3.24) until the construction of an eight-mile-long seawall in 1997. The seawall will eventually have no beach in front of it, and sand will no longer be transported past it. This loss will push the erosion problem further south, down the beaches of the Matagorda Peninsula. The first seawall will lead to another seawall, which will lead to more seawalls: a global truth. Once erosion is under control, storm-vulnerable buildings by the hundreds will quickly appear: an American truth.

Bounded by Aransas Pass (the Texan term for inlet) on the north and Brazos Santiago Pass on the south, only a few miles from the Mexican border, Padre Island is the longest barrier island in the world (batik 3.9). The Padre Island National Seashore encompasses more than half of the island's 135-mile length, but at the south end is South Padre Island, the closest thing to Miami Beach in Texas.

FIGURE 3.24
The shoreline at Sargent Beach, Texas, where the erosion rate, at the time of this 1995 photo, was sixty to one hundred feet per year. The shoreline is made up of eroding marsh mud with little sand. The sand supply has been lost as a result of human alterations of the mouth of the Brazos River, forty miles away. Sargent Beach now has an eight-mile-long seawall.

THE AMERICAN BARRIER ISLAND SCENE

BATIK 3.9
Padre Island, Texas, 1998, 87″ x 35″
At 135 miles, Padre Island, backed by
Laguna Madre, is the longest barrier island
in the world, so long that there is a climate
difference from one end of the island to the
other. The climate change affects island
shape. Rainfall is highest at the north end,
where dunes are large and well stabilized
by vegetation. At the south end, sparse
plant growth leads to smaller, more
mobile dunes.

Artist's note: Padre Island includes a 110-
mile stretch of national seashore along the
Texas Gulf shore. A combination of six
topographic maps from the U.S. Geologic
Survey and the National Ocean and
Atmospheric Administration, the design
shows our country's longest and one of the
least-developed barrier islands. The colors
are that of Easter, as I was dyeing the silk
at the same time that I was dyeing eggs
with my daughters.

The first explorer to walk on Padre Island, in 1519, was Alonso Alvarez de Pineda, who named it La Isla Blanca. Since then, the island has at various times been named Isla de Corpus Christi, Isla del Brazo de Santiago, and the mouthful Isla de San Carlos de los Malaguitas. It was called Isla Santiago when Padre Jose Balli, eventually the island's namesake, claimed the island in 1800. He built Rancho Santa Cruz de Buena Vista and a mission intended to save the Karankawa Indians, at a site twenty-six miles north of the southern tip of the island. The ranch flourished until the 1844 hurricane that caused most inhabitants to leave the island permanently. Rancho Santa Cruz, now called the lost city of Padre Island, was exhumed by moving sand dunes in 1931.

Laguna Madre, behind Padre Island, is long and narrow like most of the lagoons behind the barrier islands of Texas because the mainland shoreline follows the contours of the 120,000-year-old Pleistocene barrier island chain, stranded on the coastal plain. Laguna Madre is saltier than normal seawater along much of its length because of the high rate of evaporation in the semiarid climate and the long distance between inlets that supply fresh seawater. The wind, rather than the pull of the moon, controls the tides in Laguna Madre.

The island ranges in width from three miles to a few tens of feet. Its great length results from the small amount of water that is available to exchange between the sea and the lagoon. Tides are small, and only small volumes of fresh water flow to the lagoon from the mainland. The Island is so long that a change of climate occurs along its length. The northern portion is heavily vegetated by grasses that stabilize the dunes, some of which are fifty feet high. From mid–Padre Island south, rainfall is slight, vegetation is sparse, and the dunes are mobile, changing their shape and location according to the whims of the wind. South Padre Island is actually migrating slowly landward, while the rest of the island is stabilized by a high sand supply.

Padre Island reacts to natural forces according to its own logic. Penetration of the island by storm waves tends to occur repeatedly at the same easily identifiable locations, called overwash passes, in contrast to the situation on most North American barrier islands, where the location of the next storm overwash is anyone's guess. Some of the Padre overwash passes are old inlets, and some will become inlets in the future. In addition, large portions of the lagoon side of the island consist of sand flats sloping almost imperceptibly into Laguna Madre. On Padre, much of the island widening during migration occurs by the growth of these wind flats into the lagoon, a unique process on North American barriers, where widening usually occurs by storm overwash. These flats exist because of the lack of marshes on the lagoon side of the island. Marshes are absent because of the high salt content of Laguna Madre and the irregular highs and lows of the wind-controlled tides. Grain by grain, the island slowly migrates into Laguna Madre.

COMMON GROUND

Although often referred to as a single chain, the U.S. barriers are in reality a series of island chains interspersed with spits, like the Matagorda Peninsula in Texas, and stretches of mainland coasts, like Myrtle Beach, South Carolina. Most are coastal plain islands, but there are also delta islands on the Mississippi Delta, composite islands in Georgia, and man-made islands such as Fort Lauderdale, Florida. There are transgressive islands like Dauphin Island, Alabama, and regressive islands like Bogue Banks, North Carolina. Padre Island, Texas, is a hot dog, and Folly Beach, South Carolina, is a drumstick.

Despite the huge variety among the barrier islands of North America, they do have many strands of commonality. Most have a small sand supply, growing ever smaller because of various kinds of shoreline engineering, and most are in the earliest stages of the island migration process in response to the latest sea level rise. Many of them migrated to their present location from a more seaward location on the continental shelf. With few exceptions, including parks and wildlife refuges, barrier islands are under intensive development pressure.

Another shared element is the impact of the most recent Pleistocene (120,000-year-old?) shoreline on the modern barrier islands. The width of lagoons—e.g., wide Pamlico Sound and narrow Laguna Madre—is often controlled by the location of the old island chain. Sometimes the ancient and modern island chains coincide, as in Palm Beach, Florida, or almost coincide, as in the Georgia islands.

The most important shared element among the American islands, however, is their history—sea level history in particular. This history is why extensive barrier island development has occurred here. During the last three or four thousand years—the last half of the Holocene—the sea level has risen, flooded the mainland, and formed the embayed coast that is the necessary precursor of island formation. The difference between this coast and the coastal plain coasts of China and Brazil, where few islands exist, is almost certainly a different sea level history. These coasts have a less embayed or irregular shape, since sea level fell a few feet during the same time period. A dropping sea level will tend to straighten a shoreline, thus creating an unfavorable situation for barrier island formation.

The future of the American barrier islands promises to be most interesting, especially because the sea level is rising at the same time that development is increasing at a fast pace. Something will have to give, perhaps in two to three generations. If sea level rise continues, and especially if it accelerates, the time will come to abandon defense of the islands in favor of the salvation of our major coastal cities.

Barrier Islands and Human Realities

Awash in Politics

CHAPTER 4

THE WEST FRISIAN ISLANDS: THE DUTCH ALPS

For centuries, the North Sea waters off Holland's West Frisian Islands have been among the most dangerous to mariners of any in
the world. Buoys and lights have long steered the unwary mariner
away from the shoals and tidal deltas of the barrier island chain. Nonetheless, shipwrecks abound on the sea floor just off the Frisians. In October
1799, the British frigate *Lutine* sank inexplicably in nearshore waters, losing all hands. Originally a French warship, the *Lutine* had been captured by
the British and had fought as part of Admiral Nelson's fleet five years earlier. On this particular mission, the *Lutine* was acting as a cargo vessel, carrying a large amount of gold and silver from Great Britain to Hamburg.
No one knows why the ship drifted so far off course. A northwesterly gale
blew that night, but the storm was not one of extraordinary force. There
were rumors that the ship's officers may not have been alert due to "riotous
celebrations" the night before leaving port.

The *Lutine* sank off Vlieland, one of the five-island West Frisian chain.
Fishers later recovered a small portion of the treasure, but the shifting
about of the thick nearshore sand blanket has kept the wreck and a million
pounds worth of treasure hidden. The Lloyds of London payoff for the disaster was among the largest in the company's history at the time.

In 1857, after a storm, the ship's eighty-pound bell was discovered and
purchased by Lloyds. The bell became the famous "Lutine bell" that to this
day rings out in the giant London headquarters of Lloyds to signal word of
the sinking of another vessel somewhere in the world. On September 14,

2001, the bell was rung twice, at the beginning and end of a minute of silence observed by four thousand people in the Lloyds of London building to honor the World Trade Center victims of 9/11.

The West Frisian Islands face the tumultuous North Sea and are part of a continuous arc-shaped chain of barrier islands (fig. 4.1) connected to the German East Frisian and German-Danish North Frisian Islands. From east to west the main Dutch islands are Texel, Vlieland, Terschelling, Ameland, and Schiermonnikoog—plus three or four other small, uninhabited ones (batik 4.1). To a large extent, the evolution of these barrier islands has taken place within recorded history. When the Romans occupied the Netherlands, there were no barrier islands. The islands formed over a period of several hundred years beginning around the end of the Roman occupation (A.D. 200–300).

It is not certain where the Frisians, the local inhabitants, originated, but they arrived about 50 B.C. For much of the Middle Ages they enjoyed a

FIGURE 4.1

Map of the Frisian Island chain off the Netherlands and Germany.

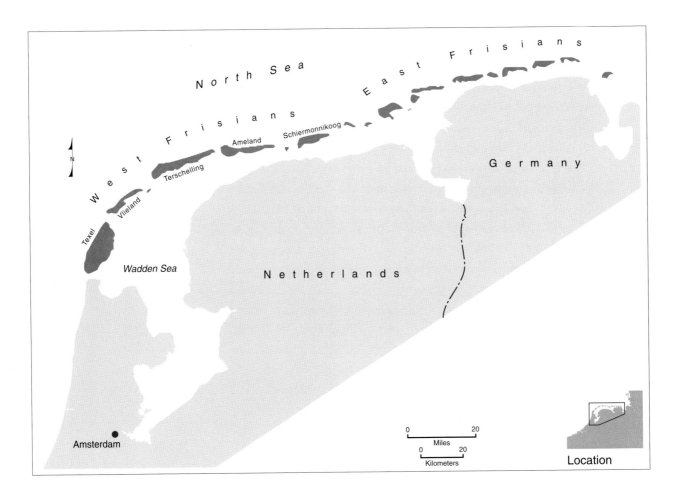

BARRIER ISLANDS AND HUMAN REALITIES

great deal of freedom but stayed on the sidelines of world events. The Romans, Vikings, Franks, Normans, Spaniards, French, and Dutch successively found these seafaring people to be very tough and difficult to rule. The Frisian language is still spoken and recently has experienced a revival. At least two of the islands, Terschelling and Ameland, were independent countries for a time. Schiermonnikoog, "island of the gray monks," was a monk colony for centuries.

One strange event occurred in January 1794 when the Dutch war fleet found itself frozen in ice in the inlet between Texel Island and the mainland. A French cavalry unit happened by, charged across the ice, and captured the fleet. A local tourist brochure dryly notes that this was "one of the few instances in naval warfare when horsemen have been decisive."

The Frisians differ from the islands of North America and most other coastal plain islands around the world in that they did not march across the seafloor to their present location, during the last of the Holocene sea level

BATIK 4.1

Dutch Frisian Islands, 1999, 31" x 44"
The Dutch island chain is unusual because it formed entirely within historic time, after the Romans left Holland. The islands were created when storms, aided by sea level rise, broke through a line of large sand dunes adjacent to the shoreline.

Artist's note: Dutch Frisian Islands is dyed to resemble the brilliant hues of a Vermeer painting. The bright yellow islands remind me of the tulips in my garden. Matt Stutz supplied maps from the Netherlands. Water movement between the mainland and the barriers creates a complex system of sandbank shallows apparent in the batik.

rise. Instead, they formed in place (fig. 4.2). Sea level rise, combined with a period of much increased storminess, allowed the sea to break through a wide band of high dunes fronting the North Sea, flooding the mainland to form the Waddensee (Wadden Sea) behind the islands. Texel Island, the first in the chain, may have formed only eight hundred years ago. Since their formation, all the islands have evolved into drumstick shapes (with the exception of Texel)—widest at their east or updrift ends (fig. 4.3) and slimmer to the west, at their downdrift ends.

The range between high and low tide is around five feet on the largest island in the chain, the thirteen-mile-long Texel Island. The tidal range gradually increases to nearly double that at the end of the East Frisian chain in Germany. Sea level is assumed to be rising at a rate of around one foot per century, the same as in much of North America. During one period, Ameland Island was subsiding at ten times that rate because of nearby ex-

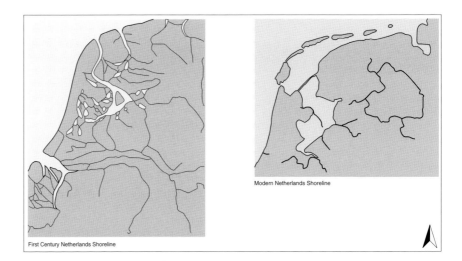

FIGURE 4.2
Maps comparing the ancient and modern coast of northern Holland. The entire Frisian Island chain probably formed within the last eight hundred years.

FIGURE 4.3
Map of the West Frisians showing the location of the dikes behind the islands that provide space for agriculture. The natural islands, without the dikes, would clearly be much narrower. For example, most of the area of Texel and Ameland Islands is new land, enclosed by dikes. Diking resulted in the loss of salt marsh on the lagoon sides of the islands.

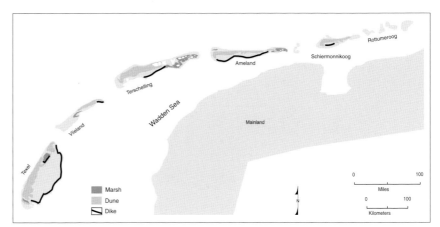

BARRIER ISLANDS AND HUMAN REALITIES

traction of natural gas. When it became evident that removing natural gas was causing sediment layers to compact and the land surface to subside, gas production was halted. In a curious natural example of accidental international cooperation, the Dutch islands are supplying the German islands with sand because of the strong longshore transport of sand from east to west.

The wide, high band of dunes that rim the ocean still exists today, although in a more landward location as a result of shoreline erosion. In Holland, which is among the flattest of all countries, the thirty-foot to one-hundred-foot dunes are major topographic features, grand and imposing in their own way, leading to their designation, somewhat derisively perhaps, as the Dutch Alps.

The Dutch recognize that vegetation is critical in dune development, but there can be too much of a good thing. Non-native grasses flourish, crowding out native species. The solution is cows—giant Scottish Highland cows, capable of thriving in the cold and windy North Sea winter climate of the Dutch islands. The cows are effective because they have selective appetites attuned to the very plants that the Dutch want to eliminate. Overgrazing is never a problem, as the cow numbers are closely watched. The polders, land areas below sea level, provide rich farming tracts (fig. 4.4). A special breed of sheep known as Texel sheep is raised on the polders. Local legend has it that the polder soil is so rich, the sheep need only stand still to graze. By the time they have chewed their last bite of grass, the blades will have grown back.

FIGURE 4.4
Ameland Island showing a cultivated forest and the rich pasture lands within the polders. Legend has it that the pastures are so rich that sheep need only to stand in one place. The grass will grow back during the time it takes them to chew the last bite! *Courtesy of the Rijkswaterstaat Survey Department, multimedia.*

Dutch geologists who have studied the islands, such as Saskia Jerlesma, believe that no islands exist on the southern Dutch coast, closer to the mouth of the Rhine River, because the sediment supply is too large. If barrier islands did form, their lagoons would quickly fill in with river sediment. Lagoons behind barrier islands all over the world are gradually filling in, but the rising sea level is working against this trend by enlarging the lagoons.

Compared to the rest of the Dutch coast, the Frisians are downright pristine (fig. 4.5), however humans have had a major impact on them. Today the Dutch chain comprises five main islands, but at one time there were many others. Today's islands were made by closing the inlets and diking the gaps between smaller islands, mostly in the twelfth to thirteenth centuries. Ameland, for example, was once three separate islands. Texel was enlarged in 1630 by connecting it to Eierland, island of the eggs—so named because of its huge number of nesting birds. After the dikes were built, the Dutch closed the gaps with artificial dunes. About the same time that big islands were being made out of little islands, land was reclaimed on the backsides of all of the five major islands. The polders, enclosed by dikes, more than doubled their size and removed the natural salt-marsh rim that once was part of the lagoon shorelines.

The inlets are smaller or narrower now than they once were, because the area of the Waddensee lagoon behind the islands has decreased as a result of diking. Since the lagoon area is decreased, the amount of water (the

FIGURE 4.5
The "sandhook" on Ameland Island, the Netherlands. This large and very dynamic feature is probably formed by waves bending around the nearby ebb tidal delta that cause longshore transport directions to reverse and move west for a short stretch of shoreline at the updrift end the island. Ameland, like all of the five main Dutch islands, is a drumstick island, fattest at its eastern or updrift end. *Courtesy of the Rijkswaterstaat Survey Department, multimedia.*

tidal prism) going in and coming out of the inlet is reduced. Less water leads to smaller inlet cross sections. Because of the construction of polders and the combining of islands, both the East and the West Frisian Islands are larger now than they ever were in the past.

The North Frisian Islands are a different story. These north–south-trending islands have less than a tenth of the area they are estimated to have had in the thirteenth century, and high erosion rates continue to this day (fig. 4.6).

From the standpoint of coastal management and living flexibly with nature, the Dutch have drawn an unwise line in the sand, and the West Frisian barrier Islands will not be allowed to retreat beyond that line. This governmental edict was issued despite the fact that there are no imperiled buildings; the Dutch consider the jeopardized dunes to be a rare natural environment, worthy of erosion protection. Another explanation for the Dutch insistence upon absolute shoreline stability is that they have the most engineered shoreline in the world. Engineering is second nature to the Dutch society. In Holland nature will be allowed to follow its course, but with the firm guidance of engineers behind it.

The good news is that the Dutch propose to hold their barrier islands in place by "soft" means, pumping sand to the beaches and by sometimes building up the dunes. Although a few rock groins are already in place on the islands, the Dutch insist that hard stabilization—rock and concrete sea-walls and the like—will not be used in the future. The beaches on virtually all of the West Frisian Islands have had sand added to some extent, and the sand pumping effort will continue at intervals ranging from two to five years, at great economic cost to the country.

Even though a line in the sand has been drawn, and is to be held by engineering means, the Dutch may ultimately have barrier islands closer to a pristine state than those in Portugal—in spite of the fact that the Portugal islands are entirely within a national park.

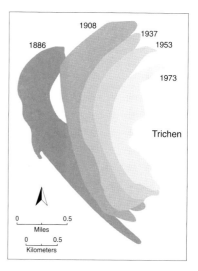

FIGURE 4.6
Line drawing showing the recent history of Trichen Island, one of the north Frisian Islands off the German coast. The cause of this rapid island change and migration is perhaps a combination of sand supply loss and sea level rise. *Modified after Dieter Kelletat.*

THE ALGARVE ISLANDS: ENEMIES ON ALL SIDES

On January 1, 2001, the barrier island chain in south Portugal (Algarve Province) entered the fourth millennium in which it played a role in Western history. Several centuries before Christ, the Phoenicians, and later the Carthaginians, sailed through its inlets to use the harbors found in the lee of the islands. For five hundred years, the Roman town of Ossanoba, at the site of the present town of Faro, was a busy port, also shielded by the islands. Starting early in the eighth century and ending in the twelfth, the Moors ruled. Later still, Prince Henry the Navigator and Vasco da Gama passed along the Algarve coast, launching the great age of discovery that made Portugal the center of the Western world.

The Moors and the navigators who followed wrote about the problems of living with dynamic barrier islands and inlets. Forts or castles, built atop high mainland bluffs directly across from inlets as part of an anti-invasion shield, were often rendered useless. Inlets migrated away, leaving the cannons far out of range of the inlet's new location. This problem became particularly troublesome in the fifteenth and sixteenth centuries when pirates streamed through the inlets, pillaged towns, and carried citizens away to slavery. In response, temporary, movable encampments began guarding the coast, replacing the imposing but static forts.

The ever-active inlets posed a navigational problem too. Lighthouses constructed on the east side of inlets soon fell, while those built to the west became uselessly marooned in the center of an island. At one point, one of the king's engineers recommended mounting the lights on wagons and moving them each year apace with the inlet's movements.

The barrier islands of south Portugal are found in the broad passage between Europe and Africa that leads to the Straits of Gibraltar and the Mediterranean Sea. They line the rim of a submerged platform or rock ledge that extends outward from Portugal's hilly southern coast, about thirty feet below sea level. Very likely the islands exist because the waves rolling toward the Straits of Gibraltar from the open Atlantic produce strong, west-to-east-flowing longshore currents. The sand flows with the currents from the eroding cliffs at the west end of the island chain, east toward the Spanish border.

The chain consists of five islands and two peninsulas along a thirty-mile stretch of coast (fig. 4.7). From west to east, they are Faro (peninsula), Barreta, Culatra, Armona, Tavira, Cabanas, and Cacelas (peninsula). Together they make up the seaward boundary of Ria Formosa Natural Park, the equivalent of a U.S. national seashore.

A number of things are different about the Portuguese islands. For one, the vertical range between low and high tide is seven feet, but that doubles during the twice-a-month spring tides. This is an extraordinarily large difference. Relative to land area at normal mid-tide, the islands lose on average between 50 and 60 percent of their area at spring high tide. Cabanas Island almost disappears.

Second, because this is a curved barrier island chain (fig. 4.8), each island and inlet has a different orientation relative to winds, waves, and currents, and that results in widely differing processes of evolution. The islands illustrate yet again the principle that no two barrier islands are the same. Armona inlet, between Armona and Culatra Islands, expands and contracts but stays mostly in one place. Most of the other inlets migrate from west to east. Fuzeta inlet, between Armona and Tavira Islands, seems to migrate or "jump" to the east after storms. Usually the new Fuzeta inlet successively occupies the next overwash pass adjacent to the former inlet. This behavior caused one building to "jump" from one island to another. The

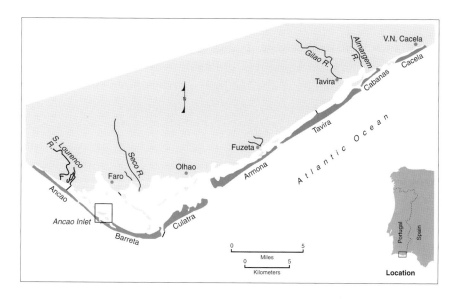

FIGURE 4.7 (above)
Map of the Portuguese barrier island chain.

FIGURE 4.8
Barretta (foreground), Culatra, and Armona Islands. Sand flows in the longshore currents from west to east or from the bottom of the photo to the top. As a result of the curvature of the Portugal island chain, clearly visible in this photo, each island has a different orientation relative to waves, winds, and storms, resulting in barrier islands with widely varying sand supplies and processes of evolution. *Photo courtesy of Joao Dias.*

ruins of an old Portuguese Rescue Service house recently moved over to Armona from Tavira Island when a storm caused Fuzeta inlet to re-form to the east of the structure. Cabanas inlet re-forms regularly after storms completely overwash and even inundate the entire island. When the storm passes, a number of inlet-like channels are created, and one of them becomes the new inlet a few tidal cycles later.

Wave energy here is high by most standards. Normal fair-weather waves are usually three to five feet high, while storm waves may be sixteen feet or higher. The islands are covered with dune grasses and a few bushes, but no trees except for those planted by the homeowners in settled areas.

As one can easily imagine, the monthly intrusion of salt water into the island interiors creates problems with the vegetation in the flats between dunes. Few land plants survive (fig. 4.9), and those represent a very restricted range of species in the areas of inundation. The groundwater is affected by this intrusion as well. The brief introduction of salt water contaminates much of the groundwater, severely limiting the fresh well-water supply available for the villages.

Most of the islands have a frontal dune ridge backed by a gently sloping surface, leading to a flat salt marsh to the lagoon. The gentle landward slope represents an apron of overwash sediment deposited before the dune ridge formed. The monthly flood tides carry sand into the island by floating it. Large amounts of floating sand and shells collect on the surface of the slowly incoming lobe of spring tide water, held on top of the water by its surface tension (fig. 4.10).

The islands are very slowly migrating toward the mainland. The island backbarrier widening during the migration process is usually accomplished by storm overwash, but Tavira and Armona Islands widen by the incorporation or attachment of old flood tidal deltas (fig. 4.11).

FIGURE 4.9 *(left)*
This low-elevation area on the lagoon side of Tavira Island is covered by the spring tide salt water several times during the month. As a result, the low elevation sand flats on the Portuguese barriers have very sparse plant covers with low diversity.

FIGURE 4.10
Sand floating on the surface of the water intruding into Culatra Island during a spring tide. Sand drops out as the tide ebbs. A floating shell is visible in the middle of the photo. It is possible that this is a significant source of sand for the Portuguese islands, helping to raise their overall elevation.

BARRIER ISLANDS AND HUMAN REALITIES

On a 1991 trip to the islands to observe island flooding during the perigean spring high tides, the highest of the year, we put ashore at the fishing village on Culatra Island. For a few days every month this village loses its soccer field to tidal flooding (fig. 4.12). Two old men sitting in chairs on the edge of a concrete slab greeted us. When we returned several hours later they were still in their chairs and the waterline from the now receding high tide had come within inches of their resting place. The old men had come to the sea's edge to see the extraordinarily high tides and knew exactly how high they would be.

Our understanding of Portugal barrier islands is based on field map and aerial photo studies supplemented by discussions with local elders. My

FIGURE 4.11 *(above)*
Tavira Island, in the Portuguese island chain, is widening on the lagoon side of islands by the incorporation of tidal deltas. The strange pattern of dunes and spring tide channels shown here is believed to reflect former channel patterns on the now emerged flood tidal deltas. *Photo courtesy of the Geological Survey of Portugal.*

FIGURE 4.12
Soccer field at a fishing village on Culatra Island. At least two or three times each month, the field is flooded—all part of life on a low-elevation barrier island where there is a six- to seven-foot difference between spring and normal tide ranges.

experience has been that the quality of memories of local island dwellers around the world is bimodal, of two kinds. Recollections are either sharp and often accompanied by an amazing intuition for how the islands evolve or they are dreadfully bad, seemingly unrelated to the reality shown in old aerial photos and maps.

Bill Neal and I worked during several field seasons with two geologists, Jose Monteiro and João Dias, from the Geological Survey of Portugal. We worked long but pleasurable hours on the islands, walking their entire lengths and vibracoring in a number of locations. When we crossed the surface of the marsh we always followed the paths made by fishers as they walked to their designated clam beds. The paths meandered a lot, and one of our team, heavily laden with coring equipment, decided to take a shorter route. Three steps off the fishers' path and he was thigh deep in mud.

In our work we used small boats belonging to other government agencies, and one craft was particularly memorable. The Portuguese tradition of superb navigation and boat handling had not been passed down to our boatman that year. We crashed almost every time we docked, and we often sped past small fishing skiffs, oblivious to the waving fists of their occupants, who were angry about the wake of our boat. I remember many times looking back to see them hanging on to the gunwales for dear life as their skiff bounced wildly. Once, when we were in a rented fishing boat in poor condition (which we overloaded with our equipment), the boat's rear end suddenly dropped off while we were under way, and down we went, luckily in wading-depth water.

There is no better example than Portugal of coexistence problems between a barrier island national park and the local citizenry. The country's populace as a whole wants the islands to remain in a pristine state. But local people already have well-established routines of fishing, building construction, waste disposal, shoreline stabilization, and recreation that inevitably clash with park goals. The interest groups that have the most impact on the management of the Portuguese barrier islands are the squatters, the port authority, the fishers, industries responsible for pollution, and those who armor the shoreline to protect their buildings.

The squatters. All told, there are 2,100 illegal houses in the Ria Formosa National Park, for the most part built by well-off outsiders, among them some prominent politicians, seeking a vacation cottage. The squatters form communities of several dozen to several hundred buildings on Faro, Culatra, and Armona Islands. At least one five-story building on Faro is built on land not owned by the structure's owner. On Armona and Faro, the squatters have even built beach-destroying seawalls to protect property that is not legally theirs. It is a perpetual Portuguese political issue; what to do with freeloaders on public lands all over the country? Newly elected politicians come in, promising to eject squatters, and they make a few well-publicized forays to bulldoze down some buildings. But it always ends in irresolution.

Once, my colleagues Bill Neal and João Dias were gathering samples on Culatra Island when an entire village streamed out to meet them, suspecting that they were government agents about to condemn their illegally sited houses. Joao and Bill quickly departed, abandoning a box of sand samples.

The Port Authority. The Portuguese Port Authority, that nation's equivalent of the U.S. Army Corps of Engineers, works completely autonomously. Portugal's park service first learned that the authority planned to build a jetty between Tavira and Cabanas Islands when the construction equipment arrived at the site, along with barges filled with big rocks. On Cabanas the jetties, located downdrift from Tavira Island, have caused shoreline retreat that necessitated construction of a seawall on the island. As erosion resulting from sand loss moves downdrift along the length of the island, the seawall is periodically lengthened. The authority seems to be concerned with navigation problems only and is planning more and larger jetties. The impact of its activities on the erosion rates of the national park barrier islands is not a consideration.

The fishers. The Portuguese fishing industry may be the least wasteful in the world. Hardly a scrap of the catch is discarded. Somebody, somewhere, eats almost everything harvested in Portuguese nets and traps. But the inshore fishing industry in Portugal, which supports thousands of low-income fishers, is highly vulnerable to coastal change, and change is a fact of life on barrier islands.

Migrating inlets change the steaming distances to docking facilities, wipe out navigation channels in lagoons, and open new lagoonal areas to the hazards of large waves coming through the inlet from the open ocean. As inlets move, the area of ocean water influence moves as well. Lagoon water salt content changes, fish and clams move accordingly, and as a result the nature of the seafood harvest changes. Whenever such changes occur, cries arise to return the inlet to its former location and restore lagoons to their former state. Nobody likes change. The effort to keep things as they are is one reason for building jetties or constructing artificial inlets. Anything to preserve the status quo. Anything to hold everything in nature perfectly still.

If the Portuguese islands were allowed to evolve unimpeded, overwash sand would cover clam beds. The fact that new clam beds will form somewhere else to replace the old ones is beside the point. To prevent overwash, the park service now builds up low sections of the frontal dune line with dredged sand and mud from the lagoon behind. Dredging the lagoon kills organisms, releases mud, and changes circulation patterns. When muddy sand is dumped on a barrier island, where it doesn't belong naturally, it produces mud-hardened dunes and releases mud to the open sea.

The polluters. Pollution behind Faro Island has increased apace with Portugal's rapidly burgeoning tourist industry and industrial development. As Ancao Inlet moved to the east, natural flushing of the polluted lagoon waters became less and less efficient. The park service gave in to pressure from a number of advocacy groups and in 1999 moved both Ancao Inlet (fig. 4.13) and Fuzeta Inlet back to the west. As the new inlets were opened by dredging, the old inlets (batik 4.2) immediately closed naturally. It was a rare but good example of island management based on scientific studies of the islands. The park service argued correctly that it was accelerating what nature would probably render in coming storms. On the other hand, the inlets would never have been moved artificially if pollution from nearby towns had been controlled.

FIGURE 4.13
The new Ancao Inlet, one year after opening (by bulldozer and dynamite). The inlet was opened in a new location to "improve" the circulation of the lagoon and within a year had formed this well-developed tidal delta. The ebb tidal delta (offshore) is very small in spite of the high tidal range because of the high waves that prevent sand from accumulating on the seaward side of the inlet. Batik 4.2 is the tidal delta from Old Ancao Inlet, which closed naturally in a few months after this new inlet was opened. *Photo courtesy of Joao Dias and Ramon Gonzalez.*

BARRIER ISLANDS AND HUMAN REALITIES

Shoreline armoring. At the updrift or western end of the Algarve barrier island chain, rapidly eroding cliffs of a red sandstone are probably the single most important source of sand. The cliffs retreat so fast that they are a hazard to people on the beach. A few years ago a tourist was killed by a falling segment of the cliff. Valle do Lobo, a large development of cottages, golf courses, swimming pools, and tennis courts, was built close to the cliff's edge, and the inevitable happened. Structures closest to the cliff fell into the sea, so a seawall was built. The wall did retard the cliff erosion for a while but without cliff retreat there is no sand for the downdrift islands. No sand means greater erosion rates.

The future of the Portugal barrier islands is fraught with problems of intergovernmental cooperation, lack of political will or power on the part of those who enforce the rules, and rapidly increasing pressure on the barrier islands from all sorts of directions. Some Portuguese officials have told us

BATIK 4.2
Portugal's Ancao Tidal Delta, 1998,
33" x 47"
The batik represents a spring low-tide view of the ebb tidal delta of Ancao inlet on the Portuguese island chain. This inlet no longer exists. It closed up naturally when a new Ancao inlet was formed artificially to the west.

Artist's note: Portugal's Ancao Tidal Delta is based on a sepia-colored slide from João Dias of the University of Algarve. The silk is an embossed floral, dyed to match a small Portuguese plate in my studio. I listened to the soulful blues of the land, called Fado, while waxing and dyeing the flooded delta.

that the relatively strong environmental status of the European Union may ultimately save the barriers, but there is some question as to whether the current political status of the islands will allow their preservation for future generations. It could be much worse—which is the lesson of Taiwan.

Portugal and Holland once were global powers that explored the vast unknown oceans of the world and opened trade routes that included contact with Taiwan, the third member of our trilogy. The Dutch actually occupied a small part of Taiwan for a brief period but were driven out by the mainland dynasty.

TAIWAN: ONLY THE MEMORY REMAINS

Just north of Wai Sun Ding Island, Taiwan's largest remaining barrier island, lies the Yun Lin industrial park construction site. Eventually, the park will extend completely across the shoreline and hundreds of yards out to sea, to be protected by massive seawalls. The park is certain to eliminate most sand transport to the south and starve Wai Sun Ding to death.

A modern research group with state-of-the-art laboratory and oceanographic ship facilities is studying the environmental impact of the industrial park on Wai Sun Ding Island. Their wave tank, a device that simulates

BATIK 4.3
Taiwan's Ancient Barriers, 2000, 36" x 48"
The faint lines along the coast of Taiwan represent the fragile islands as they were before World War II. Of thirty-five original barrier islands, only five remain today.

Artist's note: Our trip to Taiwan in February 2000 took us to three major universities. Pilkey and I met with geologists from tip to tip of the Asian country and went miles out to sea to visit barren black-earth beaches. This batik illustrates the island chain facing China, as it was before World War II, and is taken from a satellite composite. Tsung Yi Lin's introductions guided our mission.

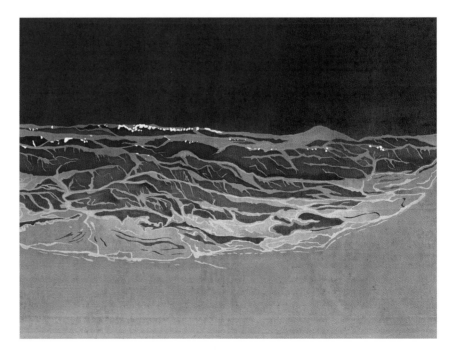

BARRIER ISLANDS AND HUMAN REALITIES

ocean conditions, is the second largest in the world. The effort is impressive and well funded, entirely concentrated on the fate of a barrier island. The problem is that the impact studies are just beginning and the industrial park is almost completed.

Fifty years ago as many as thirty-five barrier islands formed a natural chain along Taiwan's west central coastal plain coast (batik 4.3). The islands probably formed initially as spits that extended from the mainland, isolated by inlet formation during storms. Today, perhaps five genuine islands remain, and they are all vanishing (fig. 4.14).

To understand the fate of this nation's barrier islands, one must view their destruction in a national context. Taiwan (formerly called Formosa) was a backwater rather than a participant in the great cultural, scientific, and military events that unfolded over three millennia on the China mainland. Chinese began settling here in the seventh century, quickly displacing the "aborigines" of Malaysian descent. In the seventeenth century, because the Dutch and Portuguese seemed to be taking an interest in the island, the Manchu Dynasty took it over. Almost three hundred years later, in 1895, the Manchu were replaced by Japan. At the close of World War II Japan had to abandon Taiwan when the island was restored to China. In 1949 Generalissimo Chiang Kai-shek and his army fled to Taiwan after their defeat by Mao Tse-tung and the Communists. A difficult four decades of martial law and huge military expenditures ensued, with the ever-present threat of assault from Mainland China, which lay a mere hundred miles across the Taiwan Strait.

Nevertheless, Taiwan pursued industrialization with a vengeance. Because most of the country is mountainous, industry was crowded onto the

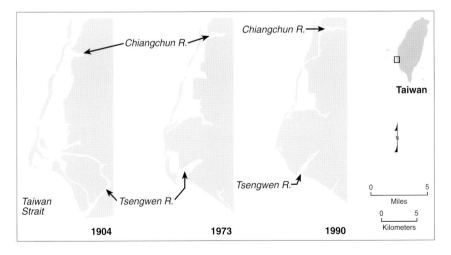

FIGURE 4.14
Maps showing the recent history of barrier islands along a section of Taiwan's west coast. Out of perhaps thirty-five original islands at the end of the nineteenth century, five remain. Most disappeared after World War II, as a result of the filling in of lagoons. The remaining lagoon (Chi Ku), which is now smaller than shown in this 1999 diagram, is behind Wan Tzu Liao Island and is threatened by a proposed industrial park, which will extend across the ocean shoreline.

small coastal plain bordering the west coast of the island, the same coastal plain bordered by the barrier islands. The coastal land was seen as a convenient place for development. The lagoons could be diked, and food could be grown and pollutants dispersed in the waters. The barrier islands would serve to protect the lagoons.

With the constant threat of Red China, keen ambitions for material success, but no credible environmental movement, the Taiwanese disregarded the health and welfare of their barrier islands. Preservation was not allowed to interfere with "progress."

It is not that these islands have actually disappeared. On the contrary, most of them are still in the same places where they were. It's just that they are no longer islands. On the ocean side, the former islands are armored with elephantine engineering structures that have completely halted all natural beach processes. In fact, the beaches are gone. On the lagoon side, the water is also gone, replaced by spiderweb patterns of dikes for fish farms and salt ponds and elevated sites that are used for industry or pig, poultry, and sugarcane farming.

The Wai Sun Ding Barriers

Protruding at a strange forty-five-degree angle from the natural trend of the mainland shoreline at the mouth of the Peigang Shi River is the largest remaining barrier island off the coast of Taiwan. It actually consists of three barriers: Tong San, Bo Tzu Liao, and Wai Sun Ding Islands, separated by small, meandering streams rather than by true inlets. The overall length of all three islands is eleven miles, and their width ranges from several tens of yards to three-quarters of a mile. Tidal range here is on the order of eight feet.

Talk about dynamic—this island may hold some sort of "land speed" record. Much of the island is said to have been eroding at a rate of 250 feet per year in recent years. A couple of lines of evidence make this astounding number believable. One is that the large steel light tower, used as a navigation guide on Wai Sun Ding Island has been moved or replaced six times since World War II. This beats the story of the Outlaw family house at Nags Head, North Carolina, which has been moved 600 feet in five separate moves over the past hundred years.

The complete lack of vegetation is a second and fundamental indication of ongoing dynamic change (fig. 4.15). On the southern end of Wai Sun Ding, near the tower, we saw nary a stalk of salt marsh nor a blade of dune grass. The Wai Sun Ding barrier island is in a temperate climate, and attempts have been made by local government to introduce several kinds of

plants, to no avail. The explanation must be the high frequency of over-wash, the consistently strong winds carrying both sand and salt spray from the surf zone, and most important, the rapid rate of island migration.

Wai Sun Ding has a single beach ridge, behind which is an undulating low-elevation surface, often flooded by storm surges. The sand is dark in color and very uniform in size, with a low shell content. Debris, composed mostly of bamboo poles from the oyster rafts that crowd the bay, virtually covers the ridge.

Two or three low points across the island bear signs of recent occupation by water that flowed across the island. While crossing one such dry channel, we suddenly realized that we were about to witness the formation of a new inlet across Wai Sun Ding (fig. 4.16). As the tide rose, the swash of the surf zone propelled ever further across the island, until finally the water in back of the island was united with that in front. On most U.S. East Coast barrier islands, witnesses to formation of new inlets are usually called hurricane survivors.

On this visit to the Wai Sun Ding Islands, I am accompanied by artist Mary Edna Fraser and Tsung Yi Lin, a former Duke University graduate student. The island can be reached only by the strange boats of the Taiwan in-shore fishing fleet. Their hulls, made of plastic pipes bent up and lashed together at one end to form the bow, were in former times made from bundles of bamboo stalks: more evidence that we live in the plastic age.

On our way out to Wai Sun Ding, as we leave the entrance to How-Mei-Liao boat harbor, we witness the process of barrier island destruction. Lining the shore of the embayment behind Wai Sun Ding Island—not a full-fledged lagoon because it is wide open to ocean waves from the west

FIGURE 4.15 *(left)*
Wai Sun Ding Island has almost no plants on it because it is migrating at a rapid rate. Low dunes that occupy the ocean side of the island are frequently covered by bamboo poles from oyster rafts and plastic debris from Mainland China that floats across the Taiwan Straits.

FIGURE 4.16
Mary Edna Fraser *(upraised arms)* celebrates the opening of a new inlet on Wai Sun Ding Island. In this high tide photo, the waters from the ocean and the waters from the lagoon are within inches of joining, which happened seconds after the photo was taken. Wai Sun Ding Island is uniformly low and susceptible to new inlet formation at many locations along its length. The survival of this island is deemed important because it shelters oyster rafts in the lagoon behind it. Consequently, there is a good chance that it will be heavily armored in the near future.

and southwest—is a string of small, one- to two-mile-long islands. They were once barrier islands. One of them, the one-and-a-half-mile-long Futzo Island, is in the process of being armored, Taiwanese style.

The saga of Futzo Island is further proof of the old adage about coastal engineering that has been mentioned earlier: once you start you can't stop. About ten years ago, the Taiwanese constructed offshore breakwaters made of large boulders along the southernmost half mile of Futzo Island. The breakwaters slowed down but did not stop the shoreline from retreating. At the same time they cut off the supply of sand to adjacent shorelines to the north, which induced rapid erosion on the unarmored portion of the island. The next response was to engineer heavier armoring along the shoreline, which was already spiked with breakwaters (fig. 4.17). The construction, meant to extend to the remaining unarmored mile of island, was under way as we watched.

Armoring proceeded in stages. First, a large concrete seawall was built, wide enough at the top for vehicular traffic. Next, the face of the seawall was armored with large concrete tetrapods. Finally, a large boulder seawall called a revetment protected the base of the line of tetrapods. This construction project amounted to an astounding piece of armoring, mightier than the Galveston, Texas, seawall that was erected after six thousand people lost their lives in the 1900 hurricane. The great Futzo seawall, stronger and more imposing than any open-ocean seawall in North America, protects perhaps a half dozen small buildings and some fishponds.

FIGURE 4.17
A massive shoreline armoring project under construction on Futzo Island. The fortifications here consist of a concrete seawall with a road on top, faced with giant concrete tetrapods, below which is a boulder revetment and seaward of which are boulder offshore breakwaters. This astounding level of fortification is protecting half a dozen buildings in the filled-in lagoon behind the island. This is the means by which Taiwan has destroyed thirty out of thirty-five of her original barrier islands and will soon destroy those that remain.

Wan-Tzu-Liao Island

The maritime forest is an artificial one. A dark copse of dense, close, and evenly spaced casuarina trees is the only component (fig. 4.18). But this patch of forest is being buried and killed by sand at its north end even as young trees are sprouting at the south end. Within the forest, the unpleasant odor of an egret rookery permeates the air. My presence evokes much squawking and wing flapping, and I decide to retreat back to the open dunes. Tsung Yi Lin meets me there and quietly notes that travel through this small patch of forest is problematic. It seems that old Chinese tradition holds that one's manhood is improved by the consumption of the meat of freshly killed venomous snakes. It is illegal to import such snakes to Taiwan, but naturally a small smuggling industry has arisen to meet the demand. Over the years, the snakes have been thrown overboard or hastily released on the island just as the customs officers arrive. The snakes that have been seen on this island, including cobras, are alleged to be concentrated in the forest. I certainly walk with more vigilance from that point on.

Wan Tzu Liao Island is on the north limb of Ding Tou Er, a cape lined with perhaps a dozen islands or former islands. Chi Ku lagoon, behind the island, is the last remaining lagoon along Taiwan's coastal plain coast. The tiny lagoon fragment of four to five square miles is crowded with bamboo rafts, from which dangle strings lined with growing oysters. The lagoon is threatened by still another industrial park that will, it is said, occupy "only a small corner" of the lagoon while extending across the shoreline to the continental shelf.

The potential loss of Chi Ku lagoon (fig. 4.19) has attracted protests from both the international and, at last, the growing Taiwanese environmental

FIGURE 4.18
The lagoon side of Wan Tzu Liao Island. The sand dune is expanding rapidly into the lagoon and widening the island as part of the island migration process. The forest is a planted one that is being killed by sand expansion on the upwind end shown here, but is enlarging on the downwind end (south) as new trees appear.

FIGURE 4.19
Almost all of Chi Ku Lagoon behind Wan Tzu Liao Island is filled with bamboo oyster rafts. Lines dangle below the rafts and provide habitat for oysters that will eventually be sold in local seafood markets.

movements. The planned construction particularly concerns the guardians of the black-faced spoonbill because this is the general area where the entire world population winters each year. Black-faced spoonbills number only seven hundred individuals (an increase from three hundred a decade ago). Many of the spoonbills hatch and are nurtured in the demilitarized zone between South and North Korea, which presents a delicate political situation for bird conservationists.

Wan Tzu Liao Island is four miles long, on the order of one hundred yards wide and ten to twenty feet high (fig. 4.20). Except for the patch of forest, the entire island is actively moving with the wind, which blows predominantly from the north-northwest. Sand accumulation on the island has been very successfully aided by meticulously handwoven bamboo twig sand fences. The island's gradual migration into Chi Ku lagoon occurs as the backbarrier flat is widened by windblown sand.

The beach is steep, with active erosion scarps and a considerable cover of plastic garbage, mostly refuse from upwind Mainland China. I believe it must be the most garbage-strewn beach I have ever seen. The island bears few signs of visitors and no facilities to encourage recreation or tourism.

There is no rest for these superactive islands. A new inlet opened within the two months before our visit, and a local fisher tells us it opened as a spring tide spilled across it. The channel gradually widened in subsequent weeks and became navigable for small boats.

FIGURE 4.20
A view to the north down the length of Wan Tzu Liao Island. In contrast to Wai Sun Ding Island, Wan Tzu Liao has vegetation and well-developed dunes, indicating a more stable island with a larger sand supply and a lower erosion and migration rate.

BARRIER ISLANDS AND HUMAN REALITIES

They are in our hands. Like rivers and forests, and like the fish and deer in the rivers and forests, barrier islands and their forests, dunes, and marshes depend on us for their well-being, their continued existence. Our political systems decide whether islands should evolve and change and respond to sea level rise or whether they should become inert sandbars or even disappear.

Politicians don't have it easy with barrier islands. Political processes and island processes just don't mix. Barrier Islands are places of change, and governments don't like change. Storms open up new inlets, salinities change, and fishers complain that the crabs, clams, and fish have moved to a new location. "Put the inlet back where it was" is the cry. Inlet channels wander about, and shippers ask politicians to build jetties and hold the channels still. Houses fall into the sea as islands migrate, so citizens ask that the migration be halted. Beaches disappear in front of seawalls built to protect the homes of the wealthy, and the less wealthy complain about the loss of the beach. It seems that to make one group of citizens happy politicians need to make the islands immovable concrete monoliths and to keep the rest happy they need to abandon the islands and leave them to fend for themselves.

In our trilogy of barrier island societies, the politicians of Taiwan stand at the concrete monolith end of the political spectrum. Barrier islands are viewed as convenient bars of sand upon which to build massive seawalls to prevent the loss of any land and then to fill in behind to gain new land. Preservation of any single component of the beach and barrier island natural system is not a consideration. Coastal problems will be engineered away and nature be dammed.

Although the philosophy of Taiwan seems very foreign to the American approach to coastal management, the United States once followed the same path. Coney Island, a part of New York City, once was a real island, the southernmost in the New York island chain. Over the past two centuries, the lagoon and salt marshes have been filled in and covered by high-rise city buildings.

The politicians of Portugal are at the "let nature roll on" end of the spectrum of barrier island management. At least, they would like to be. But there are too many people to please. Politicians and park officials struggle to coexist with fishers, squatters, industrialization, a growing local population, and an uncoordinated bureaucracy. The islands are ever so slowly creeping toward an engineered system designed to satisfy the maximum number and to insult the minimum number of interested parties. Nature, on the Portuguese islands, is something to appreciate and enjoy, but within reason.

Although in times past the Dutch have widened islands for farming and have joined small islands together to form big ones, they have now decided to preserve them, even to the point of restoring some lost salt marsh by breaking dikes. The Dutch have treated their barrier islands gently, allowing no buildings near the beach. But it is a society with a strong engineering orientation, since much of Holland owes its existence and its security to engineering of the shoreline. The Dutch can't quite step away. The engineering mentality is a control mentality, and it has decreed that the islands shall move no more. This edict will be accomplished by "soft" means, i.e., beach nourishment, but it does not portend well for the long-term future of Dutch islands. Barrier islands don't like to stand still when sea level is rising.

Delta Barrier Islands

CHAPTER 5

That Sinking Feeling

SOME of the great civilizations of the world have flourished on the rich soils of deltas, including those of the Nile, Rhine, Indus, Ganges-Brahmaputra, and Tigris-Euphrates Rivers. But the barrier islands that rim the delta fronts have not shared in this march of civilization. Even today, when people of the richest civilizations are willing to build homes on breathtakingly dangerous barrier island sites, the world's delta barriers remain largely undeveloped.

Accumulations of sediment at the point where rivers flow into a still body of water are called deltas. In the fifth century B.C., the Greek historian Herodotus coined the term to describe the Nile Delta, which is shaped like the Greek letter of the same name. Most, but not all, river deltas are lined with barrier islands. The mightiest of them all, the Amazon River, has no barrier islands because the river is carrying a huge load of mud with little sand. The real delta of the Amazon River is in the deep sea and not adjacent to the shoreline at the river mouth. Without sand there can be no barrier islands. Without protective barrier islands there can rarely be wide expanses of salt marshes or mangrove forests.

Almost all delta barrier islands have experienced a dramatic increase in the rates of erosion or island migration just in the last century. Natural causes, such as sea level rise and droughts in headwater regions, account for some increase in shoreline retreat, but humans are usually the problem.

The most fascinating thing about delta barrier islands is the endless number of ways in which they form and evolve. In this discussion, we begin our journey with the Mississippi Delta because we understand far more about its origin and evolution than about that of any other delta. We'll end with the Gurupi Delta in Brazil, about which little is known in a geologic sense and where the islands have not been previously recognized as part of the world's family of barrier islands.

THE MISSISSIPPI RIVER DELTA

Hernando de Soto, the Spanish captain on an expedition to find gold in the "Lands of Florida," explored the lower reaches of the mighty Mississippi River. In 1542 he happened upon a flood on the river and marveled at what a grand sight it was, water stretching to the horizon, broken only by hundreds of trees floating quietly by. Later, however, as settlers began to till the fertile soils of the huge Mississippi and Missouri floodplains, floods were no longer fascinating sights; they were natural disasters.

Gradually the Mississippi became a kept river. Under all but the most extreme flood conditions (like those of 1993) the Mississippi is now a pussycat, at least relative to the lion it used to be. People wanted power, so they dammed the Missouri and other tributary rivers, and the dams trapped sand. People wanted safe and uncomplicated access to the sea from New Orleans, so they built artificial levees along the lower Mississippi and prevented the normal dispersal of sediment-rich floodwaters into the delta. People wanted to keep the Port of New Orleans open, so they have kept the river in its old channel far longer than it naturally would have been, preventing the river channel from switching to the Atchafalaya channel. This has caused the delta lobe to extend so far seaward that the sediment load of the river is dumped into deep water on the continental slope (batik 5.1). People wanted to drill for oil, so they constructed hundreds of miles of seemingly random canals through the marshes, thus changing the water circulation that disperses sediment throughout the delta.

All of these alterations to the ways in which riverborne sediment is distributed on the Mississippi Delta inconveniently come at a time of rising sea level. The overall rate of sea level rise along the delta is a subject of some debate, but it probably averages around four feet per century, with lots of local variations. Because of sediment loss and sea level rise, the marshes of Louisiana, which constitute 40 percent of the nation's coastal wetlands, are disappearing at a very rapid rate. Barrier islands, which are naturally very dynamic, have become even more dynamic, and some chains (e.g., Isles Dernieres) are due for destruction within a decade or

BATIK 5.1
Mississippi River Delta, 1993, 83" x 36"
The seaward tip of the modern delta lobe of the Mississippi River shown here extends well beyond the continental shelf and spews out its sediment load far from the marshes that need it. This lobe would have long ago been abandoned in favor of the Atchafalaya River mouth except for upstream engineering activities.

Artist's note: The delta on the Gulf of Mexico extends the land seaward in the photograph that Eros Data Center provided for this batik. It was taken on the Space Shuttle Mission 61-A, October 30–November 30, 1985. The 1993 Mississippi floods had recently occurred when I made the piece, and I empathized with the victims, envisioning the delta as a bleeding heart. Every day this river delivers tons of sediment to its mouth.

two. The loss of islands will only worsen the wetland loss problem as open-ocean waves eat away the vulnerable marshes.

The Mississippi Delta is the land of the Cajuns, whose food and music are imitated and celebrated internationally. Famous for a "let the good times roll" outlook on life, the Cajuns originated in Acadia, now Nova Scotia, where, as French Catholics, they were declared unwelcome by Great Britain. They began to straggle into Louisiana in the 1750s, eventually joining an amazing assemblage of nationalities on the lower delta, including Yugoslavs, Chinese, Spanish, Italian, French, Irish, Malays, Germans, Cubans, and many freed slaves of African origin.

There were four significant settlements on the delta barrier islands in those early days. Isle Dernieres, then called Last Island, was the first tourist resort on the Louisiana coast. By 1850 it was thriving and even had two small hotels, but on Sunday, August 10, 1856, the Last Island Hurricane killed 140 people and destroyed every building on the island except a single small hut. That was the end of civilization anywhere on the Isles Dernieres chain.

Grand Terre Island developed into a center for privateers in the early nineteenth century, complete with a flourishing stolen-goods market in New Orleans. Jean Lafitte, the most famous of the pirates, even rented warehouses in New Orleans to store his booty before sale. The U.S. Navy captured the island and halted the piracy in 1814. The island of Cheniere Caminada, just west of Grand Isle, had 250 homes and 1,471 people in 1890. Eight hundred residents perished in the 1893 hurricane.

Grand Isle, a slightly higher and more stable island just west of Grand Terre, has been more fortunate than the other barrier islands in the group. First settled in the early 1800s the island had a flourishing fishing industry, and with the help of large numbers of slaves, it produced cotton and sugarcane. After the Civil War it became a resort community, to which people from New Orleans flocked to escape the malarial mosquitoes. Now widely known as the Cajun Bahamas or the Cajun Riviera, Grand Isle is an awkward collection of beach cottages, docks for fishing boats, tourist watering holes, a summer population of twelve thousand, and a serious erosion problem (fig. 5.1). It has a growing boulder seawall, and the beach has been nourished no fewer than twelve times. The mayor of Grand Isle, in a move deserving of some sort of creativity award, offered the opportunity for anyone to have his or her name chiseled into a seawall boulder in exchange for paying the cost of getting the boulder to the beach.

FIGURE 5.1
Looking west over Grand Isle, Louisiana, the only remaining populated Mississippi Delta barrier island. In the nineteenth century at least three towns on other barrier islands along the delta were destroyed. *Photo by Greg Stone.*

In short, the experience of settlements on these islands has been loss of property, loss of life, abandonment, or a costly struggle to hold back the sea. One should not ignore history on a delta!

The Mississippi Delta (fig. 5.2, batik 5.2) began building about seven thousand years ago when the Holocene sea level rise slowed down and the level of the sea was close to its present position. The Mississippi River dumps its load in the Gulf of Mexico, by building out lobes or tongues of sediment extending across the continental shelf. Periodically—every couple of thousand years—the lobes are abandoned and the river chooses a new and shorter course to the sea with a steeper gradient, leading to more efficient flow. During the last seven thousand years, the Mississippi has occupied five lobes (fig. 5.3) and abandoned each in turn (table 5.1). The present-day Mississippi River Delta lobe would like to abandon its channel in favor of a new one, down the Atchafalaya River west of the current river mouth. Only the U.S. Army Corps of Engineers stands in the way of complete abandonment of the present river mouth.

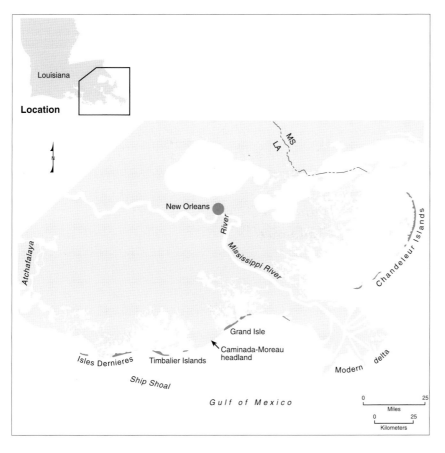

FIGURE 5.2
Map of the Mississippi Delta showing the principal barrier island chains.

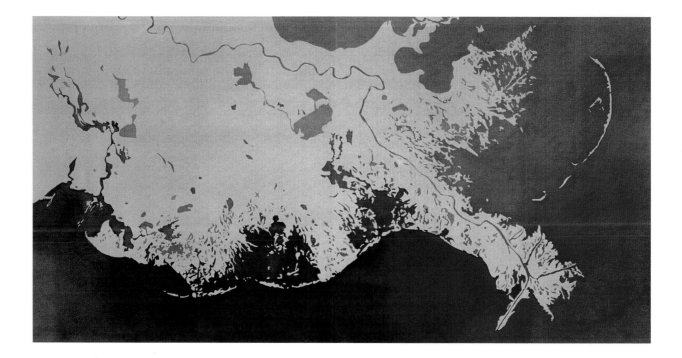

BATIK 5.2
Louisiana's Disappearing Chains, 1997,
33" x 65"
Five barrier island chains, each of a
different age, line the bird's-foot delta of
the Mississippi. Each of the chains formed
immediately after a delta lobe was
abandoned, when the river found a new,
shorter, and more efficient route (steeper
gradient) to the sea.

*Artist's note: Louisiana's Disappearing
Chains* is interesting in that these islands
are sinking out of sight in my lifetime. I
intend the watery coastline to feel like lace
and the delicate island chains to be akin
to necklaces surrounding the Mississippi's
bird-foot delta. The dyes on silk merge two
gradations of colors, which amplify the sea
level rise. Numerous Louisiana Geological
Survey maps that indicate the shifts of the
islands through the decades are the
foundation of the art.

While a Mississippi River Delta lobe is active and building out to sea, no barrier islands can form, but after the river mouth chooses a new location and the old lobe is abandoned, the creation of barrier islands begins. This is the beginning of stage 1, according to Louisiana geologist Shea Penland, who first recognized that the delta barrier islands go through a three-stage cycle from birth to death.

Local sea level change and the infusion of sand are the driving forces behind the changes. Sea level rise rate varies on the delta, depending on the rate of compaction of delta sediment. The greater the rate of sea level rise or the smaller the supply of sand, the more quickly the islands go through the three-stage cycle.

Stage 1. Without new sediment, rapid shoreline retreat of the abandoned delta begins. As the breaking waves remove the easily suspended mud, sand is left behind to be pushed ashore and washed up onto the lip of the eroding lobe, eventually to be concentrated into an actively migrating barrier island. Under the never-ending assault of waves, some of this sand is carried laterally by longshore currents to form new barrier islands on the flanks of the former delta lobe. This process is very much like the spit mode of formation of coastal plain barrier islands, where instead of a delta

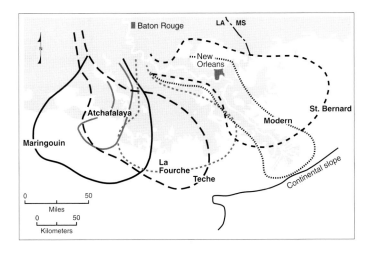

FIGURE 5.3
Outlines of the five delta lobes that make up the Mississippi Delta. The oldest is the Maragouin Lobe, which began perhaps 7,500 years ago and was abandoned 5,000 years ago (table 5.1). The youngest is the Atchafalaya Lobe, which is not yet fully occupied by the river.

TABLE 5.1 **Names and Ages of the Six Principal Delta Lobes of the Mississippi River Delta**

(B.P. is Before Present)

Name of Delta Lobe	*Age*
Maringouin	7,500 to 5,000 years B.P.
Teche	5,500 to 3,800 years B.P.
Saint Bernard	4,000 to 2,000 years B.P.
LaFourche	2,800 to 800 years B.P.
Balize	1,000 years B.P. to present
Atchafalaya	present to –

lobe, a protruding former ridge on the coastal plain is eroded to supply sand to the adjacent barriers.

The Caminada-Moreau headland, formed from the LaRouche delta lobe (table 5.1) and abandoned eight hundred years ago, is at stage 1 (fig. 5.3). The headland itself is now marked by a narrow barrier island, frequently overwashed by storms, retreating at more than 110 feet per year, and backed by a large salt marsh. Sand flows from the headland, west to the Timbalier Islands and east to the Caminada spit and Grand Isle. The islands both to the west and to the east are rapidly disintegrating because the sand supply furnished from the headland is insufficient to maintain them.

Stage 2. The next stage of delta island evolution is separation of the flanking barrier islands from their headland and formation of an unattached arc-shaped barrier island chain. Sand, once obtained from the eroding headland

119

and transported away by longshore currents, is now obtained from the continental shelf and shoreface, washed ashore by waves. There are two stage 2 barrier island arcs on the delta today. These are the Isles Dernieres (fig. 5.4) just west of the Timbalier Islands and the Chandeleur Islands on the east side of the modern delta (fig. 5.1). The Dernieres were formed six hundred to eight hundred years ago during an early retreat phase of the LaRouche lobe. The Chandeleurs are the oldest islands on the Mississippi Delta. They formed on the rim of the Saint Bernard Lobe more than two thousand years ago and may have migrated as much as six miles landward from their original position.

Both the Chandeleurs and the Dernieres are migrating in a landward direction, and both are deteriorating, but the Isles Dernieres are doing it in

FIGURE 5.4
The Isles Dernieres after Hurricane Andrew in 1992. These islands are frequently breached and overwashed during storms. Between storms the islands are partially repaired. *Photo by Greg Stone.*

much more rapid fashion because local sea level rise is faster. The Isles Dernieres have lost 75 percent of their land area in historic time (fig. 5.5). For the Isles Dernieres, the change from a headland with flanking barrier islands (stage 1) to a barrier island arc (stage 2) occurred just within the last 150 years.

Stage 3. The final event in the Mississippi island cycle is the complete loss of the islands, which become an underwater shoal or sandbar. Seaward of the Dernieres chain is Ship Shoal, believed to be the barrier island chain that once formed at the rim of the Maringouin delta lobe, the oldest on the delta. The shoal is more than thirty miles in length and continues to migrate landward at rates between twenty-five and fifty feet per year. Stage 3 is reached when there simply isn't enough sand to keep an island going.

For better or worse, the United States, just like the Dutch on the West Frisian Islands, has decided to hold the shoreline in one place. On any re-treating coast this is a very costly proposition, and the one in Louisiana is particularly so. Sums of up to $60 billion per decade are rolled about on the tongues of Louisiana politicians—federal money, of course.

If nothing is done, the islands will eventually disappear and waves will attack the marshes, salinities will change, fish populations will migrate, and oil well platforms will be exposed to waves. On the plus side, new barrier islands will form on the margins of the marshes, an event that seems to be

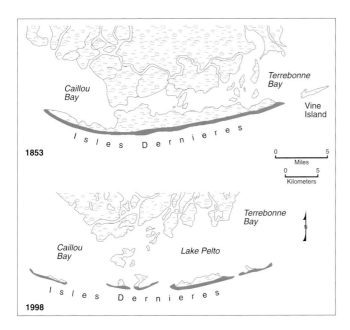

FIGURE 5.5
The deterioration of the Isles Dernieres between 1853 and 1998. Although the Dernieres are expected to disappear as part of the normal cycle of the Mississippi barriers, the process has been accelerated by human activities that have reduced the sand supply to the islands.

occurring already. All of these things will happen anyway; human activities over the last century have just sped things up a bit.

Plans are afoot to allow Mississippi floodwaters to spill through the levees beyond New Orleans to bring in much-needed sediment. But most of that sediment will simply settle in the canals crisscrossing the marsh rather than nourish the marshes. Seawalls have already found use on East Timbalier Island to protect oil production facilities, and offshore breakwaters have been emplaced on Raccoon Island in the Isles Dernieres. Can these measures hold the islands in place for decades to come? Timbalier Island already has its third generation of seawalls. The islands moved back from the walls, which sank rapidly and became offshore rocky ridges, now barely visible above the waves (fig. 12.18). Some marshes are being "protected" by small dikes along their rims with floodgates to control water flow. Is a diked marsh really a marsh so far as its use by the larvae of marine organisms is concerned? It is not at all clear that spending large sums of money will solve the delta's problems.

One political element in the societal debate is the distress of the residents of Terrebonne Parish over the potential loss of much of their land area. Their concern is understandable, perhaps, but only a few thousand people will be directly affected by land loss here, compared to a few million people on the Nile Delta protected by the eroding barrier islands. The parish is already 48 percent land and 52 percent water. Another concern is that the loss of marsh area will increase the likelihood that New Orleans could be flooded in a big storm. Would it be better to fix that problem by raising the dikes around the city?

The societal debate will undoubtedly continue, and very large amounts of money will be expended to hold the line in the mud. In all likelihood the three-step life cycle of the barrier islands of Louisiana has come to a halt. Humans, using concrete and sand, will determine how the barrier islands will evolve from now on. Time, tide, sea level rise, and politics will determine the success or failure of these efforts and the future of the Mississippi Delta.

THE NIGER DELTA

Oil is king on the Niger Delta, and life in the small villages on the barrier islands has been enormously affected. The chief source of cooking firewood for the women of the Ibani tribe of Bonny Island used to be nearby mangrove trees. Now they are afraid to use mangrove wood. It is too dangerous, they say, since the wood sparks and even explodes as the residue from years of oil spills catches fire. So they are forced to travel long distances to get other types of wood.

Residents of the fishing village of Oloma on Bonny used to get back to their island from the mainland via a small tidal creek that wandered through the mangroves. Now, however, the creek has been dredged, filled with pipelines, and declared off-limits for use by local people. During the rainy season, when the sea becomes very rough, villagers must go back and forth to the island by a different route, exposed to dangerous wave conditions. The Niger Delta barrier islands can be tough places to live.

Bonny Town (population eight thousand in 1996) on Bonny Island provides a typical people history of a Nigerian barrier island. Well before the coming of the Europeans, the Ibani tribesmen on the barrier islands had a thriving trade with the Ibo tribe in the interior. Fish and salt were traded for livestock and vegetables. When the Europeans came it was only natural that Bonny Town would become a center for two-way commerce between Europeans and the Ibos, a trade that soon gravitated to a lucrative traffic in slaves. Palm oil replaced the slave trade in the 1800s, and prosperity reigned for the rest of the nineteenth century. By the 1930s, however, the role of Bonny Island in international trade was gone, displaced by Port Harcourt, fifty miles up the Bonny River. Only the small subsistence fishing villages on the barrier islands prospered.

Then along came petroleum. Oil transfer terminals were built beginning in the 1960s, and a liquefied natural gas plant on Bonny Island, which began operation in 1999, is one of the largest industrial plants on the African continent. The problem is that schooling opportunities often did not reach the remote barrier island villages, so local people on Bonny Island did not get the high-paying jobs. Many of them must travel to towns in the hinterland for work, leaving their families behind.

The fishers in the small villages have been most damaged by the march of industrial progress. Chronic oil spills—some large, many small, most the result of carelessness and lack of government oversight and some the result of sabotage—have reduced both fish and shellfish populations. Nets and floats are destroyed by the coming and going of small oil rig supply boats (sea trucks) that sometime overturn dugout canoes, especially at night. Groves of coconut, banana, mango, and other trees have been removed to make way for petroleum production infrastructure. Canals, pipelines, pumping stations, and gas flares at night all add up to a highly altered and increasingly uncomfortable living environment.

The Niger River Delta is a wedge of sediment that has been accumulating for at least sixty million years (fig. 5.6, batik 5.3). The sediment pile forms an amazingly symmetrical half circle, 28,000 square miles in area, that protrudes 150 miles into the Atlantic Ocean along Africa's equatorial West Coast. Among the world's largest deltas, with a population of seven million people, it is said to be the breadbasket of all Nigeria and is endowed with immense natural resources, the most prominent of which is oil.

In the delta's upper reaches, the Niger splits into the Nun and Forcados Rivers, which then branch off into a number of other rivers (distributaries) before reaching the sea. Twenty barrier islands rim the seaward edge of the delta, where each inlet is a river branch (fig. 5.7). The westernmost inlet is called the Benin River, and the easternmost is the Imo River.

Counting the barrier islands to the west, beyond the delta proper, there are twenty-eight islands, with connected sand supplies, along 343 miles of Nigerian shoreline. They form the second longest barrier island chain in the world.

As in the case of the Portugal barrier islands, the curved coast results not only in different wave orientations striking the shore but also in different wave heights and different directions of longshore sand transport. On average the waves are largest (greater than four feet) on the western half of

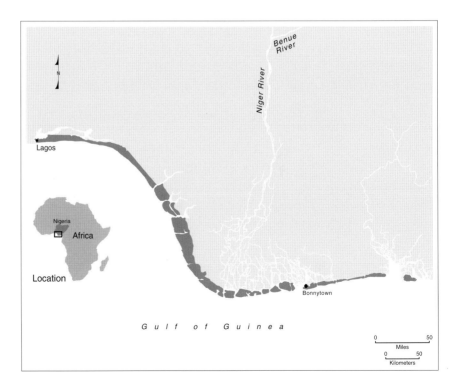

FIGURE 5.6
Map of the Niger River Delta.

DELTA BARRIER ISLANDS

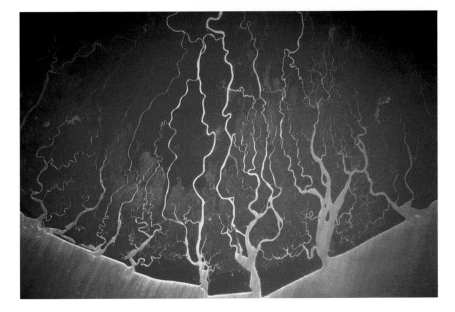

BATIK 5.3 *(above)*
Niger River's African Coast, 1999, 65" x 37"
The unusually wide islands (three miles) lining the tropical Niger Delta are rapidly eroding, mostly because of the loss of sand supply from the river. Dams and canals are the principal causes of the sand loss. At 343 miles, the Niger barrier island chain is the world's second longest.

Artist's note: The Niger River batik is dyed the colors of Nigerian weavings. The design derives from Russian maps and tactical piloting charts. The art in my mind is likened to an ancient Mother Earth, who is praying at the continent's edge. You can see her red eyelash and pregnant outline in the islands on the Niger's delta.

FIGURE 5.7
Six islands along the rim of the Niger Delta. Note their extraordinary width relative to the island length, which reflects a long period of island accretion in former times when the Niger River carried a large sediment load. *Photo by NASA, scientific data purchase image.*

the delta and less than three feet to the east. Tides range from four to eight feet, increasing in an easterly direction across the delta. On the eastern half of the delta, where the islands form distinct "drumsticks," longshore transport is primarily to the east. On the western half, the longshore currents reverse and most beach sediment moves west.

Behind the islands, a vast 3,600-square-mile mangrove forest, the third largest in the world, covers the delta. Most of the Niger Delta barrier islands are ten to twelve miles long and one and a half to three miles wide, with maximum elevations of ten to twelve feet. Their extraordinary width reflects a high sand supply and a long period of seaward widening by the addition of successive gentle low sand ridges (beach ridges) parallel to the shoreline. These beautiful islands are covered by tropical rain forest and have broad beaches lined with coconut palms and colorful fishing villages. By all accounts the barrier islands of Nigeria are among the loveliest in the world.

At the present moment in geologic time, most of the fresh water (and hence most of the sediment) is coming through the inlets on the western half of the delta. As a result, the barrier islands along the eastern half of the delta are sand-starved and have been going through a destructional phase. Those on the western half, with a larger sand supply, are in a constructional or island-widening phase. Within a few decades or centuries, the situation can be expected to reverse if humans don't interfere (a most faint hope). The river will change course and empty through the inlets on the eastern delta half, and the constructional-destructional phases of the islands will reverse.

In just the last few decades, this idyllic situation of alternating seaward widening and moderately eroding barrier island shorelines has changed. Thousands of years of island widening have come to a screeching halt. Most beaches, on both the western and the eastern delta fronts, are retreating at rates of between 50 and 100 feet a year. In a 1996 report, Nigerian geologist A. Chidi Ibe noted annual erosion rates (in east to west order) on the Niger Delta of 55 to 75 feet at Escravos, 60 to 70 feet at Forcados, 45 to 65 feet at Kulama, 50 to 60 feet at Brass, 60 to 75 feet at Bonny, and 35 to 45 feet at Imo. Very dynamic numbers indeed. Under these circumstances, a typical U.S. 150-foot beachfront lot would be gone in three years!

The villains in this clash with nature in the delta are mostly the same as those on the Mississippi Delta. Leading causes of erosion include trapping of sand by dams on the rivers, trapping of sand and reduction of river flow by canal construction, and sand mining at river mouths and even on the beaches. Subsidence caused by pumping of both water and oil from beneath the delta adds the finishing touches to a human-induced debacle.

Because the islands are wide it will be many years before they actually disappear so that waves can roll into the mangrove forests behind them. Nonetheless, the super-rapid retreat of the Niger River barrier island shorelines has increased the rate and nature of coastal flooding, displaced entire villages, dislodged oil production and transport facilities, destroyed farms, and almost wiped out the beach tourist industry.

Raphael Adewoye, the director general of Nigeria's Environmental Protection Agency, says that the nation's coastal degradation is being arrested by following the nation's "action plans such as the National Policy on the Environment, Vision 2010, and the National Agenda 21." Solving the problem of the degradation of Nigeria's barrier islands, however, requires much more than absurd soothing platitudes from a bureaucrat. In fact, the task, perhaps similar in magnitude to that in the Mississippi Delta, would prove daunting even to a wealthy country. In Nigeria, a land of governmental chaos, ubiquitous dash (bribery), and immense poverty, the future of the barrier islands is recognized as a national problem only on some distant horizon. The continued degradation of the Niger Delta islands seems assured.

THE NILE DELTA

"We said we would build the Sadd El-Ali [the Aswân High Dam] and we built it–you colonialists! We built it with our own hands." So went the words sung by every Egyptian school child in the early 1970s, a celebration of the nation's immense pride in its engineering accomplishment. The people, with the help of the Soviet Union, had built a mountain of a dam, large enough to swallow up seventeen Pyramids of Cheops. Today the Egyptians are still proud of the dam, although major environmental impacts are dulling the luster of this nationalistic symbol. The dam, just like those in the Mississippi and the Niger River headwaters, has reduced the supply of sand to the coast, in this case by 90 percent.

When I was young, every American schoolchild knew that the Nile River flooded every year and that each flood provided new nutrients and fresh soil in which to grow next year's crop. This was why Herodotus wrote, "Egypt is the gift of the Nile." The floods don't happen anymore, the nutrients are gone, and manufactured fertilizer now replaces the river

BATIK 5.4 (opposite)
Nile Delta Desert Islands, 1999, 52″ x 36″
The low-lying desert islands along the Mediterranean rim of the Nile Delta have been completely cut off from any source of fresh sand by the Aswân High Dam and six thousand miles of canals on the delta. Erosion rates are high, and dramatically increased future flooding of the mainland is anticipated as a consequence.

Artist's note: Dan Stanley's explanations of the Nile Delta in his National Museum of History office at the Smithsonian Institution brought life to this batik. He defined the distinct problems of the Nile, but maps alone could not visually illustrate the information. This batik looks like a lush lotus leaf, but in truth it depicts the Nile's parched edges. The design came from an image published in Stanley's scientific paper in the 1998 *Journal of Coastal Research.*

mud. In fact, so much river water is removed that during most of the year salt water extends ten to twenty miles up the two remaining river channels.

Once, the Nile River branched out into seven distributaries as it flowed into the broad delta plain at the end of its four-thousand-mile journey across Africa (fig. 5.8, batik 5.4), and each distributary formed a separate river mouth at the Mediterranean shoreline. Land reclamation and extraction of water for irrigation long ago removed five of the seven distributaries, leaving only the Rosetta (to the west) and the Damietta today. Now wastewater and industrial pollutants are discharged in large volumes into the lagoons and not through the river mouths.

Until the beginning of this century, the Nile Delta was in a build-up mode. Entirely because of the impact of humans, the delta is now in a destructional phase (fig. 5.9), but the Aswân High Dam isn't the lone culprit. There are a number of smaller dams and six thousand miles of canals on the delta, an area smaller than the state of Delaware, with a population of forty million.

The Nile Delta islands are barrier bars that enclose four lagoons, locally called lakes. The islands are not named on maps or charts, and in local use people refer to them by the names of the adjacent lagoons, from east to west Idku, Burullus, Manzala, and Bardawil. Idku is the small lagoon a short distance west of the Rosetta river mouth, Burullus lagoon is between the Rosetta and the Damietta River mouths, and Manzala lagoon is east of the Damietta, bordering the Suez Canal at its east end. The Bardawil lagoon, which in part is a nature reserve, is east of the Suez Canal on the Sinai

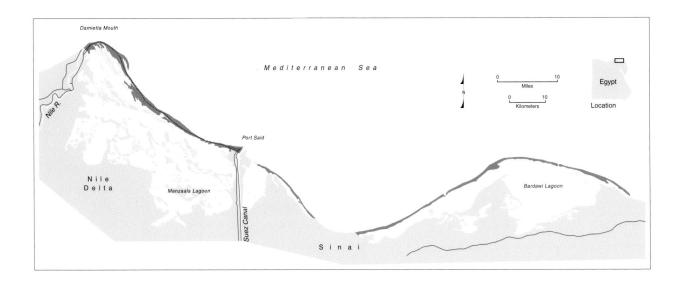

Peninsula, well off the delta. Among the Egyptian islands, the barrier islands of the Bardawil lagoon are the most remote and least altered by humans.

Nile Delta barrier islands are long, narrow, and low strips of sand, mostly remote and undeveloped, with little vegetation. Because the tides are small, generally with a range of one foot or less, the tidal water flow is small also, so the inlets, which are all jettied, are widely spaced. The islands are generally less than six feet above sea level and are occasionally overwashed by winter storms with six- or seven-foot waves. Normal winter waves are three to four feet high; in the summer they are much smaller. Small, low dunes abound, but occasionally dunes reach heights greater than thirty feet.

The islands formed as a spit of sand, extending from each of the original seven river mouths that once protruded seaward, as the Rosetta and Damietta currently do. The spits extended in an easterly direction in response to the wind and waves that come mostly from the northwest. Nile River sand on beaches extends past the Sinai Peninsula, the Gaza Strip, and all the way to Israel.

The Egyptians view their barrier islands as their main line of defense against storms and sea level rise. A thirty-mile-wide band of land next to the Nile Delta shoreline is mostly less than six feet in elevation, highly vulnerable to the whims of the sea and highly dependent on the protection of the barrier islands. The Nile Delta islands are slowly migrating landward, partly in response to delta subsidence but probably mostly because of the loss of sand from the Nile River. They are also slowly deteriorating. The island-widening component of the barrier island migration process occurs mostly when sand is blown into the lagoon, which widens the sand flat on the back of the island (as at Padre Island, Texas).

Nine developing countries, in particular Sudan and Ethiopia, have plans to make additional use of Nile water, which will further exacerbate the problems of the Egyptians living on the delta. Once the Sudanese and Ethiopians solve their serious problems of internal warfare, it is inevitable that they will turn their attention to making the desert bloom with Nile River water. This concerns many people, as it has the potential for a regional war! Egyptians believe that they already utilize every drop of Nile water and cannot tolerate any reduction of flow.

THE MEKONG DELTA

Site of continuous fighting during the Vietnam War, the Mekong Delta remains a household name in America. Although Americans can leave the war behind and move on in their daily lives, the residents of the Mekong Delta are not so lucky. The coastal strip of the delta, including the barrier islands, was one of the areas of particularly intense Agent Orange application. Between 40 and 50 percent of the mangrove forests of the Mekong Delta were destroyed by herbicides, mostly Agent Orange, during the Vietnam War. In addition, unmapped minefields, booby traps, and unexploded ammunition are still discovered on some islands, and some mainland forests remain unharvested because of these hazards.

The Mekong River starts on the Tibetan Plateau and at one point or another constitutes the borders of six countries: China, Myanmar, Laos, Thailand, Cambodia, and Vietnam. The Mekong Delta, like the Niger Delta, is the breadbasket of the rest of the country, one of the greatest rice-growing areas in all of South Asia. The Mekong annually brings a billion tons of sediment to the sea, much of it deposited near the shore, pushing the shoreline in a seaward direction, one hundred to three hundred feet per year.

The delta begins 230 miles from the South China Sea at Phnom Penh, Cambodia, where the river splits in two, to form the Tien and the Bassac. Downstream the Tien further subdivides into six main distributary channels and the Bassac into three, forming the "nine dragons" that constitute "the Mouths of the Mekong." The shoreline, which forms the base of the triangular delta, is about 350 miles long. The eleven barrier islands of the

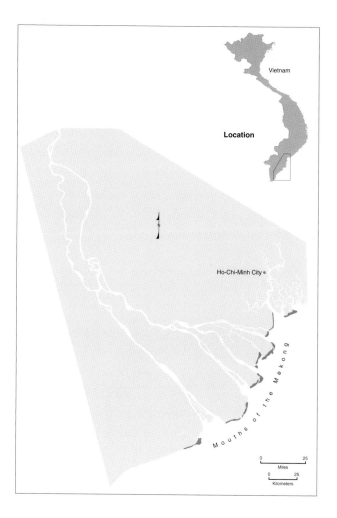

FIGURE 5.10
Map of the Mekong Delta shoreline.

Mekong Delta rim the delta lobes that separate each of the nine dragons (fig. 5.10, batik 5.5).

The islands are thin strips of sand, usually less than a few hundred yards wide, between two and fifteen miles long, rising ten to fifteen feet above the high tide line. The strong currents, produced by a tidal amplitude of ten feet and more, combined with the river flow, restrict the barrier islands to the tips of the delta lobes and prevent their extension across the river mouths. The inlets are mostly meandering tidal streams, much like those observed crossing Wai Sun Ding barrier island in Taiwan and Santa Barbara Island in Colombia.

The islands are constructed as spits by strong north to south longshore currents. At the end of every delta lobe, new island chains are under construction, charged with fresh river sand. A mangrove forest forms in the low areas, or swales, between the new spit and the previously formed spit.

BATIK 5.5
Mouths of the Mekong, Vietnam, 1998,
52" x 47"
Supplied with abundant sand, the islands on the lobes of the Mekong Delta are rapidly growing seaward. Such growth is not accomplished by island widening, as in the case of the Niger Delta, but rather by the successive addition of new islands.

Artist's note: Mouths of the Mekong reflects a sadness associated with the Vietnam War. Land mines still threaten the lives of those who remain. The colors of the tropical forest contrast with the bloody color of the South China Sea. I see this batik as a skeleton-faced Buddha with belly rolls that are irrigated rice fields. Matt Stutz researched and found U.S. military maps on which the idea is based.

In this fashion, barrier islands, curved to fit the shape of the outer delta lobe, are created in series, each to be superseded by the next, more seaward barrier island (fig. 5.11). Barrier island after barrier island (fig. 5.12) is slapped on at the lobe end. This process is in contrast to that of the Niger Delta, where the shoreline built seaward by the widening of individual islands rather than through the construction of new islands.

Once new islands form, the old islands become cheniers, high and dry sand ridges separated from older and younger islands by strips of mangrove forest. In this fashion the whole delta complex is growing seaward, leaving at the seaward end of each delta lobe a thirty-mile-wide (or more) strip of curved sand ridges. This field of alternating mangrove mudflats and sandy chenier ridges, somewhat similar to the smaller chenier fields on the west-

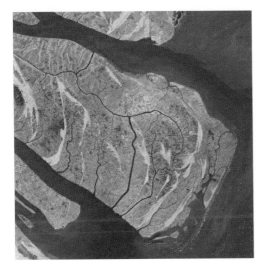

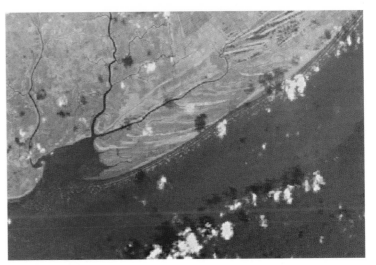

FIGURE 5.11 *(left)*

Satellite view of a single lobe of the Mekong Delta, Vietnam. The curved yellow stripes are islands and former islands (cheniers) separated by mangrove forests, shown in green. Each of the Mekong lobes is rapidly growing seaward by the addition of new barrier islands. As a result, the lowermost thirty to forty miles of each delta lobe consists of successive curved sand ridges, each a former barrier island. *Landsat Thematic Mapper (TM) image furnished by the Tropical Rainforest Information Center.*

FIGURE 5.12

A barrier island at the tip of a Mekong Delta lobe. The growth from east to west (right to left) of this island is clearly shown by the curved spits, each of which was once at the active end of the still-lengthening island. Compare this with very similar recurved spits north of Captain Sam's Inlet, S.C. (see fig. 3.11). *Landsat TM image furnished by the Tropical Rainforest Information Center.*

ern Mississippi Delta, probably formed within the last three thousand or four thousand years. The curved cheniers or former barrier islands on the lower delta lobes control the natural stream drainage, resulting in curved tributaries. Naturally, because the cheniers are elevated, they make the best sites for roads and highways, so the roads follow the same curved pattern as the nearby streams, all of which reflect what was once the arc shape of an island at the end of the delta lobe.

The sand supply of the Mekong Delta islands, unlike that of most of the world's delta barrier islands, probably remains in a relatively healthy and abundant state. Laos and Cambodia will change this situation when they construct dams on the Mekong and its tributaries, as they are expected to do.

As Vietnam begins to prosper, the remaining mangrove forests are under increasing pressure to provide charcoal, firewood, and timber and, most of all, to make way for shrimp farms and fishponds. Shrimp farms may prove to be a far larger hazard to the outer delta's environmental prosperity than Agent Orange has been. Tropical mangrove forests happen to exist under the ideal salinity, temperature, and water-depth conditions for shrimp farms. According to one study, more than half of the remaining mangrove forest of the lower Mekong Delta was removed between 1990 and 1995 and replaced by lucrative shrimp farms. It is a worldwide problem. Nations from Ecuador to Thailand are rapidly removing mangrove forests, a bonanza for the business world and a catastrophe for the marine ecosystem, for without the mangroves, larval forms of many marine organisms will have no place to call home.

The Gurupi islands are like no others. Fifty of them stretch along the 356-mile Brazilian shoreline between Marajo Bay and Sao Marcos Bay to the south. They are situated immediately south of the equator and the southern branch of the Amazon River at Belem. This coast displays a strange jagged sawtooth pattern (fig. 5.5), with the teeth of the saw being dozens of small headlands, each capped at its seaward end by small barrier islands (fig. 5.14). The Gurupi barrier islands are thin, half a mile to two miles long, ribbons of sand plastered up against mangrove-covered headlands that formed as a rising sea level invaded former river valleys. Strictly speaking, this is not an island chain, since there is no lateral connection between islands. There is no transfer of sand from island to island. All sand movement is in the on- and offshore direction, as the huge tides ebb and flow.

The Gurupi River forms the boundary between the Brazilian states of Para and Maranhao and empties into the sea approximately in the center of this shoreline stretch. A number of towns line the shore, among them Algodoal, Salinopolis, Atalaia, Ajuruteua, and Cedral. The towns are just

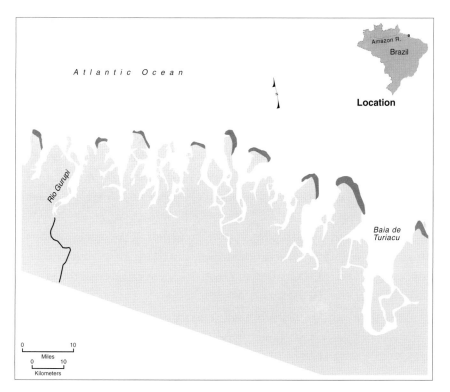

FIGURE 5.13
Map of a portion of the sixty-six-island Gurupi barrier island group on the northern Brazilian coast, south of the Amazon River mouth. Technically, this is not a true island chain, since sand is not exchanged from island to island.

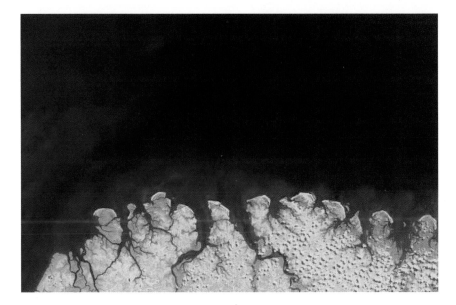

FIGURE 5.14
Landsat image of the Rio Gurupi coast
near Para Brazil. The islands are thin, light
strips at the end of each lobe. Mangrove
forests are dark green. Tidal amplitudes as
high as twenty-one feet along this coast
produce strong tidal currents that prevent
islands from extending across the mouths
of the lagoons and rivers. *Landsat 7 image
from the U.S. Geological Survey.*

behind the barrier islands with their beautiful beaches that form the basis
of a modest tourist industry in this forever warm tropical region.

Tidal amplitude is very high, perhaps the highest for any barrier island
group in the world. It ranges from seventeen feet to twenty-two feet.
The rule that barrier islands are not found where tide range is greater
than fourteen feet applies to coastal plain islands—not delta islands. Wave
energy is low because of the very wide (100 to 120 miles) continental
shelf. Immediately in front of the island chain is a several-miles-wide
band of rapidly shifting sandbars, constantly changing in location and
presenting a significant hazard for the unwary navigator. Most of the sediment of the Amazon flows out to sea or to the north, away from the Gurupi islands, but clearly some sand escapes to the south—or perhaps did
so at some point in the past, because the Gurupi islands seem to have a
large sand supply.

Little is known scientifically about these islands. They are on no one's
list of the world's barrier islands (except ours). Mary Edna Fraser's friends
"discovered" the islands as they all perused satellite imagery of the world's
coasts looking for striking scenes to batik.

The Gurupi islands (batik 5.6) are most closely related to the Mekong
Delta barrier islands. Like those of the Mekong, the Brazilian islands have
a distinct arc or crescent shape, and none of them extend even slightly into
the adjacent estuaries or bays because of the strong back-and-forth tidal

BATIK 5.6
Rio Gurupi, Brazil, 1998, 72" x 36"
The Rio Gurupi islands form a 356-mile-long chain, but strong tidal currents prevent sand exchange from island to island. These small independent islands form at the end of small delta lobes that slowly grow seaward by the addition of new islands.

Artist's note: The image was shot from a digital camera system, controlled by students participating in NASA's EarthCam project. Recognizing the barrier islands, medical imagist Jim Nicholson called excitedly when he saw the photograph. The relationship of water and land creates a wonderful interaction under the cloud layer. Pilkey was surprised to find these poorly documented barrier islands in Brazil that were unknown to him.

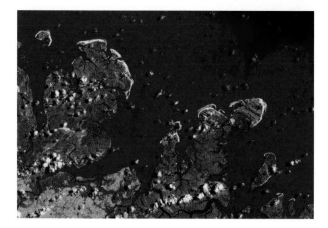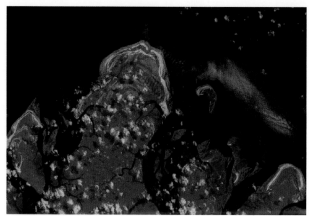

FIGURE 5.15
Satellite image of a delta lobe of a small river along the Gurupi coast of Brazil. On both sides of the lobe, longshore currents have built spits and islands whose orientations indicate sand movement landward, toward the Brazilian mainland. *Landsat TM image furnished by NASA.*

FIGURE 5.16
A satellite image of a delta lobe along the Gurupi coast that is clearly building in a seaward direction. Behind the active island at the tip of the lobe are a number of curved, light-colored strips that are former barrier islands, or cheniers, abandoned as the delta lobe prograded seaward. The barrier islands at the tip of the Gurupi Delta lobe appear to be a miniature version of the curved barrier islands at the tip of the Mekong Delta lobes. *Landsat TM image furnished by the Tropical Rainforest Information Center.*

currents (fig. 5.15). The Gurupi barrier islands, on individual delta lobes, have small inlets separating them that seem to be more like creek crossings rather than true inlets. Also like the Mekong islands, tidal deltas at barrier island inlets on the Gurupi islands are usually absent or insignificant. Still another similarity to the Vietnam islands is that the Gurupi islands seem to have built sequentially seaward (fig. 5.16). At the seaward tip of some of the Gurupi headlands is a mile-wide band of cheniers or abandoned barrier islands, now landlocked behind the active island. This is equivalent to the thirty-mile-wide Mekong Delta band of cheniers.

In huge contrast to the islands of the Mississippi Delta, which are formed as the delta lobes retreat landward, the Gurupi and Mekong Delta islands are formed as the delta lobes expand seaward.

THE SINKING ONES

Roughly a third of all barrier islands are delta islands, but global distribution of delta barrier islands is highly skewed (fig. 5.17). In the Southern Hemisphere they make up 43 percent of all islands (appendix, table 2). Fifty-seven percent of Southern Hemisphere barriers, measured by length, are coastal plain islands—including islands in Madagascar, Mozambique, and Southern Brazil. The situation is reversed north of the equator, where 24 percent of barrier island length is along the rims of deltas, compared with 76 percent for coastal plain islands (including Arctic islands).

The reason for this global disparity in island abundance, briefly discussed in chapter 2, has to do with a difference in the history of sea level change

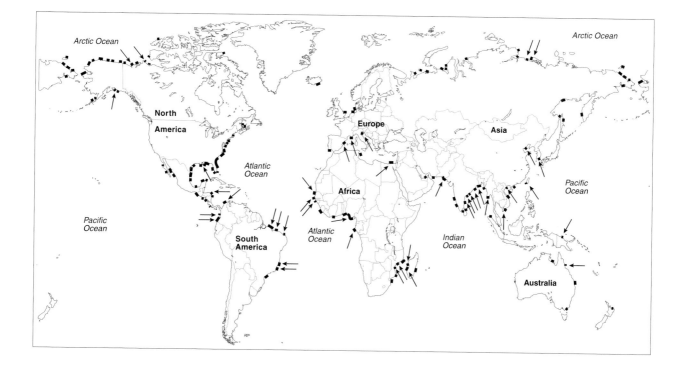

FIGURE 5.17

Barrier island chains along all of the world's coastlines. The difference between delta island abundance in the Southern and the Northern Hemispheres is probably attributable to variations in sea level history over the past five thousand to six thousand years.

in the two hemispheres over the past few thousand years. It appears that south of the equator, sea level reached a postglacial peak and then dropped by as much as twenty feet during the last three thousand to five thousand years. This reduced the size of flooded river valleys and other embayments and tended to make the shoreline less crooked relative to what it had been at peak sea level. With smaller embayments and less water to flow in and out of lagoons with the tides, conditions were less than optimal for island formation. Instead, as along much of the southern Brazil coast, lagoons were mostly enclosed by spits attached to the mainland rather than by true islands.

North of the equator, sea level rose during the same time period, an event that expanded rather than contracted lagoons. Other things being equal, larger lagoons will have larger tidal prisms. The larger flow of water in and out of the lagoons favored the breaking up of the spits that had straightened the shoreline, as nature likes to do on sandy coasts, into islands.

Delta barrier islands live in a sped-up world. Their processes of evolution, migration, widening, narrowing, and inlet formation happen quickly. Many delta island chains are in the process of self-destruction, some from natural causes, most from unnatural causes. Predictions about the future of delta islands should be expressed in decades or centuries, while the future of coastal plain barrier islands may best be viewed in the context of centuries or millennia. No wonder delta islands are the world's least developed barrier islands.

The islands that rim river deltas have two things in common: a huge supply of sediment from the river and a high rate of sea level rise as the river sediment compacts and causes the entire region to sink. Delta islands are extremely sensitive to their river sand source. Rivers everywhere are being altered to serve humankind, and human-caused changes in river discharge, and especially sediment load, translate to immediate changes in delta barrier islands. Most of these problems are twentieth century in origin, and the twenty-first century will certainly see even greater and more rapid changes in the islands as inland populations grow and more rivers are "tamed."

The responses of islands to human-induced changes also demonstrate the "workings" of barrier islands and their relationship to human activities, which mostly reduce the supply of island-building sand. Dams on the Nile River have taken sand away, which resulted in very rapid shoreline erosion rates. The extraction of water and oil from the Niger Delta has contributed to the rapid decrease in barrier island size as the sea level rise increases apace with the land's sinking. The strategy of holding delta lobes in place and preventing natural lobe switching hastens Mississippi Delta island destruction. The Gurupi and Mekong Delta islands remain "untouched" by upstream human impacts, but that will likely change during this century. Clearly the future of delta barrier island chains is change and more change. Many of these fragile bits of land will disappear in the twenty-first century.

CHAPTER 6

Colombia's Pacific Islands
A Sinking Tropical Paradise

THE GREAT TUMACO EARTHQUAKE

ONE of the greatest earthquakes of the twentieth century occurred on the Pacific coast of Colombia, South America, during the night of December 12, 1979. Later named the Great Tumaco Earthquake, after the nearby town, it measured magnitude 7.7, very similar to the August 17, 1999, earthquake near the sprawling city of Izmit in western Turkey. Unlike the Izmit quake, the Tumaco quake did not kill tens of thousands of people—but only because that remote and rugged part of the Colombian coastline is sparsely populated. Nevertheless, the tremblor was enormous, so much so that hundreds of miles away in cities from Bogotá, Colombia, to Quito, Ecuador, local residents noted that high-rise buildings swayed like reeds in a breeze. The final death count was estimated at five hundred.

The epicenter of the Tumaco earthquake was located a few miles offshore of the Colombia-Ecuador border. Five fishing villages—Curval, Timiti, Mulatos, Iscuande, and San Juan de la Costa—which lie mostly on riverbanks, were destroyed. San Juan de la Costa, which suffered the most, sat on a barrier island on the rim of the Río Patia Delta. The village was literally wiped off the face of the earth, and an estimated 250 villagers perished.

Tsunamis were the problem. Tsunamis are oceanic waves that are caused when the seafloor—in this case the adjacent continental shelf—shifts instantaneously as an earthquake strikes. The shifting seafloor acts like a paddle against the ocean water column, generating waves. These waves or sets of waves are fast moving and usually cause widespread destruction when they hit land. The famous woodblock print by Japanese artist Hokusai that portrays a tsunami wave as a scaled-up version of a normal breaking wave is probably inaccurate. The real thing is more like a plateau of water barreling toward the shoreline.

The Tumaco quake spawned three waves that struck the island at low tide. The first, according to San Juan de la Costa survivors, crashed ashore about ten minutes after the quake stopped. The third wave in the set was the largest, perhaps eighteen feet high, and it drove a five- to six-foot wall of water across the entire island. People who had fled their houses (at three in the morning) during the violently shaking quake heard a thunderous roar an instant before they were swept away into the jungle or into nearby tidal channels. To add to the terror that the villagers must have experienced, cracks appeared in the island, and in some places sand liquefied, turning temporarily into quicksand. The entire island dropped five feet in elevation. Much later, the first observers to reach the village after the tsunami were surprised that a large number of the village's pigs had mysteriously managed to survive.

Reports from the July 17, 1998, earthquake and tsunami in Papua, New Guinea, are hauntingly similar to the Great Tumaco Earthquake of 1979. A 7.1 magnitude quake produced three large waves that completely destroyed two villages, Arop and Warapu. Arop was on a barrier spit, similar in size to the devastated Colombian barrier island. Judging from debris left hanging from the trees, one of the waves was as high as a four-story building. Some survivors said the incoming wave sounded like a jet airliner taking off. More than three thousand people living in tropical lowlands very much like those in Colombia were swept away and killed.

In the weeks and months following the Tumaco earthquake, the barrier islands along a fifty-mile stretch of Colombia's coastline began to erode at an accelerated pace. The frequency of flooding by waves at high tide increased alarmingly because the islands were now at a lower elevation. Life was no longer tenable on a rapidly shrinking, frequently flooded island, and the survivors were forced to move to the mainland. A few of them attempted to return to the island, but there was so little of the island remaining—the earthquake had left its ineluctable mark—that they were forced to retreat once more to the mainland and abandon their view of the sea.

On my first trip to Colombia's Pacific coast in 1988, our team—Jaime Martinez, Juan Gonzalez, Bill Neal, and I—made the startling discovery that barrier islands existed along five hundred miles of the shoreline. Nothing in the scientific literature indicated that Colombia possessed sixty-two barrier islands (fig. 6.1). Only five of the islands bore the Spanish designation "*isla*" on navigation charts, although all of the inlets, or *bocanas*, had names.

Neither scientists nor cartographers recognized these barrier islands for what they are. Maps of the coastal regions of the world missed the islands, and lists of the great barrier island chains did not include them. One reason is that the lagoon shorelines are so poorly defined that island outlines often cannot be discerned, even with the use of aerial photos and satellite or radar imagery. From the local perspective, the barrier islands differ little from the natural levees along the banks of the rivers. Both are long, low mounds of sand, covered by rain forest and mangroves.

Five centuries of mapping and charting had not exhausted the possibilities of discovery. It was tremendously exciting.

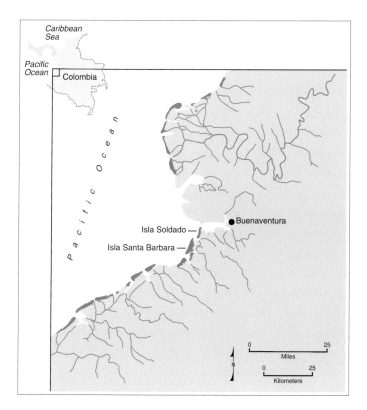

FIGURE 6.1

Map of a portion of the Pacific coast of Colombia near Buenaventura Bay.

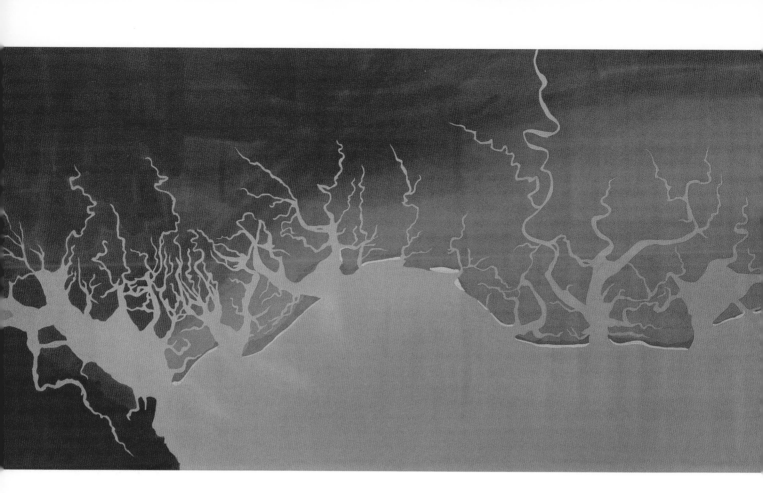

BATIK 6.1

Sinking Colombian Shores, South America, 1998, 34″ x 63″
The tropical delta barrier islands of the Pacific coast of Colombia near the entrance to Buenaventura Bay. Numerous short rivers, originating at the crest of the nearby Andes Mountains, provide a huge sand supply to these earthquake- and tsunami-prone islands.

Artist's note: When Pilkey was working on a grant from the National Geographic Society in 1995 to study the Colombian barrier islands, he came to my studio with the original field maps of the newly discovered islands. It was astonishing to me to hold these papers in my hands, knowing that even today, new knowledge is coming to light about our Earth and that I would be the first—that we know of—to illustrate these islands.

The islands on Colombia's Pacific border must be among the most dynamic anywhere (batik 6.1). They are held in one place, more or less, by the huge supply of sand that roars down the slopes of the Andes Mountains. The islands narrow when they sink after an earthquake and then widen again until the next earthquake. The flooding by waves at high tide that drove the remaining villagers away from San Juan de la Costa Island was actually the process by which the waves brought sand ashore, ever so gradually, to widen and elevate the island.

The islands behave the way they do because the Nazca oceanic plate is being forced under or is subducting under the South American continental plate. The dense Nazca plate is sliding under the edge of the continent, dragging the seafloor down with it, forming the Peru-Chile undersea trench, just offshore of the west coast of South America. The Andes mountain range is formed when the edge of the less dense South American continental plate crumples against the Nazca plate. The mountains are constantly folded, faulted, and forced upward by the oceanic-continental plate struggle. Earthquake activity is a natural result of this tectonic mountain building, but unlike the uplifting Andes mountain range, earthquakes usually sink the islands.

The high Andes mountain range, sitting within easy sight of the coast, continually erodes and supplies fresh surges of sand to the river mouths with each flood. The processes that weather out mineral grains along steep grades of the mountain streams in this high-rainfall climate guarantee a large supply of sand.

On the surface of things, Colombia's Pacific coastline should not have a long island chain. Barrier islands generally aren't found along the leading edge of a continent at the contact point of huge colliding tectonic plates. Such coasts are too steep for island construction unless deltas are present, which happens to be the case on the Colombian coast.

The accumulation of numerous small deltas that are the endpoints of many small rivers flowing out of the mountains serves to construct the Pacific shoreline of Colombia. Some of the rivers along the central regions of the coast are the Baudo, the San Juan, the Anchicaya, the Naya, the Micay, the Yurumangui, the Raposa, and the Patia. The deltas of these and many other rivers form a platform of land that extends one to five miles out from the base of the Andes and is covered by mangrove forests. This narrow platform allows islands to form on what is otherwise a very steep and forbidding coastline.

A few short gaps in the island chains occur where rock cliffs abut the shoreline or where subsidence and erosion are so rapid that the sand supply cannot keep pace. In this situation, the mainland shoreline consists of eroding, ocean-facing mangrove forests. Given a lapse in subsidence rates and time gaps between earthquakes, barrier islands will form again from sand pushed onshore and along the shore by waves.

Geologists classify the dark-colored sand found on the Colombian islands as "immature." This kind of sand makes its way to the beaches and islands when the sources are close, and the streams that feed sand to the coast are short and steep. The sand contains a wide variety of fragments from volcanic and sedimentary rocks and minerals, including abundant flakes of mica, a glassy mineral whose presence indicates that the sand has been neither weathered nor transported for long. Light reflects off the surface of each tiny, flat mica flake and gives the beach a beautiful sparkle in the sunlight. Mature sand, characteristic of beaches and islands on non-mountainous coasts, is light-colored and dominated by the mineral quartz.

Some of the local people believe that the beach sand in Colombia has therapeutic properties. We once watched as the companions of an old man buried him in beach sand as he lay prone on his back. He held an umbrella in his hand to ward off the rain from his otherwise exposed head. He claimed that the sand helped the aches and pains from his arthritis.

One unusual aspect of the Colombian barrier islands is that their evolution is not controlled or influenced by storms, at least not to the extent of those on the east coast of the United States. Very destructive storms like hurricanes do not occur in this area; they do not even pass by offshore. The waves that routinely strike the islands in fair weather and at high tide range from two to three feet in height and may reach twelve feet during rough seas. Storm surges, the scourges of hurricane-prone coastal plain coasts, are unimportant here because the narrow, steep continental shelf doesn't allow large volumes of wind-driven storm waters to pile up. The biggest wind-generated waves are probably associated with El Niño, when there is an increase in storms and the average sea level rises as much as a foot.

Because of the lack of big storms and the rarity of large waves, new inlets form quietly across the Colombian islands. They form usually during the high spring or full moon tides when the waves wash through the island, cutting small channels that widen during later high tides. This process stands in sharp contrast to the way inlets are violently and instantly created during the devastating hurricanes that ravage the east coast of the United States.

This region also has one of the highest rates of rainfall in the Western Hemisphere. The annual rainfall along the coast is on the order of two hundred inches (this is similar to the rain forests of the Olympic Peninsula in Washington State). At the headwaters on the nearby western slopes of the Andes, the annual rainfall may exceed three hundred inches.

The first report that we prepared on the barrier islands of Colombia was published in 1992. It included sketch maps of the shapes of the islands that we drew, based on aerial photographs and satellite images, as well as conventional maps and charts. Later, however, when we started our field studies, we found that our maps were outright wrong. In most cases we portrayed the islands as larger than they really are. The problem, we discovered, was the ubiquitous mangrove trees—several species can grow in the water on the fringes of the islands as well as on the island itself. We had the same problem that the original mapmakers did. They didn't recognize the islands because the mangroves, particularly the black mangroves, blurred the island boundaries so much that they were often not visible, even from planes or satellites.

The only way to determine the actual boundaries was to trudge across each island in as many places as possible until we hit water (fig. 6.2). The islands are covered by tropical rain forest, and walking across them can be adventurous and treacherous. At first, we expected deadly anacondas, boa constrictors, and other bloodthirsty beasts to prey upon us at every turn. So as we hacked our way through the jungle growth with machetes, making as much noise as possible, we felt like bold explorers, ever ready to fight off attackers. However, this approach only seemed to attract bees. Once we put our blades back into their sheaths, we relaxed as we made our cross-island forays, and they proved remarkable. The beauty of the tropical vegetation, the brightly patterned butterflies and multicolored parrots, the strange noises and the profusion of fragrances conspired to amaze us.

PEOPLE AND DYNAMIC ISLANDS

The barrier islands are remote, inaccessible, and sparsely populated by people largely descended from native Colombian Indians and African slaves who escaped and found freedom in remoteness. Few peopled communities are currently found on the islands of Colombia's Pacific coast, although most were probably settled at one time or another. Cows still graze on some islands, and wild pigs thrash through the dense vegetation on others. But time, tsunamis, and tropical vegetation have erased the memories and detritus of most of the islands' former dwellers. Tumaco (population 65,000), near the Ecuador border, and Buenaventura (population 160,000), at the head of Buenaventura Bay, are the only towns of any size along the

FIGURE 6.2
Jaime Martinez trudging through the rain forest on Santa Barbara Island. It was necessary to make frequent forays across the islands to ascertain island boundaries because island width could not be accurately determined from aerial photos.

eight-hundred-mile coast that extends from the border with Panama on the north to that with Ecuador on the south. Most barrier island and riverbank villages are populated by fewer than two hundred people.

Life is relentlessly hard. The frequent earthquakes, the tropical climate, the pollution, and the incidence of diseases like malaria combine with *bandido* and guerrilla activity and poverty to make life arduous, even in the beautiful tropical paradise that is the Colombian coast.

Where fishing villages do exist, the people tend to be wary but friendly. In 1989 my longtime friend Bill Neal, Rob Thieler, then a Duke University graduate student, and I taught a "short course" (similar to a seminar) on coastal science to a group of a dozen Colombian marine scientists. Our method of instruction was to visit several of the Colombian islands.

We arrived unannounced at a small village that was perched atop a river levee in the lagoon behind the barrier islands. Our intention was to have lunch in the local school building and learn a bit about life in the village. As it turned out, we gave most of our sandwiches to the schoolchildren as they swarmed around us. And the native folk were so pleased with our visit that they prepared a special dinner of buttered clams and rice. While the meal was certainly appetizing, I was apprehensive about it, since I had caught hepatitis once before from eating steamed clams on a field excursion in North Africa. Yet, since no one else seemed to be worried, I dared not refuse a small serving. After everyone had been served, a large portion remained and with great flourish, it was served to me. I think because of my gray beard I had been designated as "the most important" member of the group. Much to my relief in the days that followed, I found that my liver and I had survived the clams unharmed.

The story of one island village gives grim testimony to the harsh saga that is the reality of life there. Isla El Choncho (Pig Island), first settled in 1906, was so named because of the large number of domestic pigs that once roamed free. Like most communities on these barrier islands, the population was fewer than two hundred people, all of whom lived simply off the land. The villagers grew rice, avocados, yucca roots, oranges, and coconuts that supplemented their diet of pork, fish, and shellfish. The people were aware that their island was eroding, but they had learned how to live with it. The inhabitants watched as each year the beach retreated in October and November and then partially recovered from January to April. Houses threatened by eroding beaches were moved regularly, often laterally to a different part of the island. The walls of the houses were built in movable sections, designed in advance to be taken apart and moved easily and quickly (fig. 12.7). When we visited the village in 1989, the people constructed a "seawall" with twelve-inch-wide wood planks, laid end to end, in an effort to stop the full moon high tide wave swash from entering the village. Bill and I smiled to each other as we noted the contrast between erosion control that a twelve-inch-high board would provide in this modest Colombian village and the elaborate concrete or pumped-in sand barriers constructed by the U.S. Army Corps of Engineers on a wealthy American barrier island. And we also noticed that the villagers were moving one of their buildings at the same time.

Then, in November 1991, an earthquake of magnitude 6.0 struck the region centered just twenty-five miles away from El Choncho. While the island did not suffer a direct hit from a tsunami, the pattern of island erosion shifted dramatically in the months that followed. Whereas the waves previously would overwash the island two or three days each month, now they overwashed during most daily high tides. The beach stepped up its retreat to a rate of more than thirty feet per year, and the recovery part of the annual cycle of beach loss and beach recovery no longer happened.

The island had subsided perhaps 3 feet during the earthquake. The final blow came when a new inlet formed during an extreme high tide in June 1996. It soon widened to 150 feet and deepened to 50 feet. The time had come for the people to move off the island.

The new village is now located some three hundred yards away, across a tidal channel and atop a ridge of sand. Gone is the sea breeze, the sea view, the immediate access to the sea by boat, and the ability to spot approaching strangers at a distance. Gone is a way of life.

Clearly the Colombian islands are not habitable, and just as clearly people have a hard time staying away from them.

Isla Santa Barbara

As far as we could make out, a single family lived on Isla Santa Barbara (fig. 6.3). We discovered them as we were exploring by boat, wandering through the complex tidal channel system that meandered through and around the dense mangrove forests behind the island. Situated in a small clearing, their home consisted of an open platform, elevated six feet above the ground, with a thatched roof. The family members included mother and father, three children, and grandmother, or so we assumed. They owned several dugout canoes, some pigs, and a barking dog.

They welcomed our unexpected intrusion, so much so that the father of the household, with immense agility, climbed a coconut tree, and collected several coconuts for us. Then he showed us a path across the island that we had missed.

Isla Santa Barbara is one of the older islands on the Pacific coast of Colombia, if any of these islands can be considered old at all (fig. 6.4). The trees are larger on this island than on its neighbors, which is the best evidence for its longevity. The surface of the island has subtle ridges of sand, beach ridges that once marked the position of a shoreline, that run parallel to the shoreline. In addition, Santa Barbara is crisscrossed with small tidal channels, which are navigable by a skiff at the highest tides. Three of the channels actually extend to the open ocean, joining together to meet the open sea at a single point. These are not inlets—they are too small. Instead they lazily meander through the island like a bubbling brook across a meadow. The channels were likely inherited from the lagoon as the island migrated over the mangrove forest and its labyrinth of meandering tidal channels, sometime in the distant past.

Isla Santa Barbara is about five miles long, three hundred yards wide, and rises typically about six feet above the normal high tide line. Soldado Inlet separates Isla Santa Barbara from Isla Soldado in the north. Its southern end is bordered by Raposa Inlet. The Rio Raposa flows out from the Colombian mainland behind Santa Barbara and splits into two final distributaries or branches, each of which spills into the sea through the inlets on the north and south of the island.

FIGURE 6.3
The homesite of the only family that lived on Santa Barbara Island. Dugout canoes, large and small, are the principal means of transportation on the Colombian Pacific coast.

The Isla Santa Barbara tidal deltas are "horn deltas," long bars of sand, or horns, that extend seaward at least two miles from the tip of islands. They do not form the arc or fan-shaped sand body that is characteristic of most barrier island inlets around the world. Horn deltas occur in areas that experience the highest tidal ranges that barrier islands can withstand, from ten to fourteen feet. Barrier islands do not exist where the tides exceed this range—the rushing tidal currents wash them away. The exceptions to the rule are delta islands.

The horn deltas on the Colombian islands form giant breakwaters that extend offshore and partly protect the islands from waves that travel from the north or south. In front of the islands, between the horns, offshore sandbars run parallel to the shoreline and abate the energy of the waves. The bars are easy to spot as lines of breaking waves.

It became clear to us that the offshore sandbars play an important role in island evolution. At low tide, waves break on the bars hundreds of yards away from the island beaches, and the water close to the beach has the appearance of a millpond. Only at mid to high tide levels do waves break on the beach. This was true of Isla Santa Barbara and of neighboring Isla Soldado, as we would discover later.

The lack of windblown sand accumulations, or dunes, on any of the Colombian islands was a mystery to us. We certainly saw and felt sand as it was blown by the winds along the beach. Perhaps the perpetually damp

FIGURE 6.4
An aerial view of Santa Barbara Island. This island is wider and older than Soldado Island, shown in the distance. Because of its large sand supply from the nearby Andes Mountains and a relatively low subsidence rate, this regressive island has built seaward in recent decades.

nature of the sand and the dense vegetation prevented much sand from accumulating into dunes. Geologists who have studied the tropical islands of the Niger Delta also have puzzled about the lack of dunes there.

In order to unravel the history of Isla Santa Barbara, we used cores and ditches as our primary tools. As it turned out, our costly, sophisticated vibracore equipment, designed to gradually vibrate an irrigation pipe four inches in diameter into the sediment using an air compressor, failed to work. Instead we resorted to the lower-tech and more direct method of smashing the pipe into the sediment with a thirty-five-dollar sledgehammer (fig. 6.5). We were able to penetrate into the island as much as eighteen feet, a good length of core in such sandy sediment. And we were able to avoid carrying a heavy air compressor from a boat, across the surf zone, and through the jungle.

The pipe, now filled with a column of island sediment, was extracted from the ground, often with difficulty, using a block and tackle attached to a heavy tripod. The pipe was then split open, exposing the history of the island for the last few hundred or thousand years. By close examination, we hoped to distinguish beach, dune, overwash, inlet fill, tsunami, and mangrove forest deposits and to learn the sequence in which these environments had occurred. This process would determine the island's geologic history.

The trenches that we dug to supplement the information gained by coring triggered lively discussions. A path was cleared through the jungle, a

FIGURE 6.5
Sledgehammering a core tube into the sediment. After the twenty-foot irrigation pipe is forced in, block and tackle attached to the tripod is moved over the pipe to facilitate its extraction. Mangroves grow along the stream bank in the background.

four-foot-deep ditch was excavated, one wall of the excavation was carefully smoothed, and all of us convened on hands and knees at the edge of the ditch to analyze the revealed layers. Did one layer over there indicate windblown sand? Overwash? Was another a buried soil layer formed between storms or earthquakes? Was yet another a tsunami layer? Just what should a tsunami layer look like? Occasionally our discussions grew a bit heated, but we always welcomed the intellectual respites from the physical labor of ditch digging (fig. 6.6).

We learned that the layering beneath Isla Santa Barbara is fairly uncomplicated. The upper layer, three or four feet in thickness, is sand that got there by some combination of overwash, wind, and tsunamis. Beach sand, often ten feet thick, composes the next layer, and below that is mangrove mud. We interpreted this layering to indicate that Isla Santa Barbara is regressive, as discussed in chapter 2 (fig. 2.3), and has been building seaward by plastering successive layers of beach sand on the beach.

The underlying layer of mangrove mud told us something else. The mud formed in a mangrove forest, at the high tide line, behind an ancestral Isla Santa Barbara that was seaward of today's island. Since the mud is now ten to fifteen feet lower than its former intertidal position, either the sea level rose or the land sank or both.

Isla Soldado

Isla Soldado is the first island in the barrier island chain that begins at the entrance to Buenaventura Bay. Soldado, a community of two hundred people, exists on the extreme south end of Isla Soldado. The island is about three miles long, and at high tide it is only fifty yards wide over most of its length (fig. 6.7). At low tide, the exposed width of the island expands to three times that size and reveals a wide expanse of beach. Most of the island stands about four to six feet above the normal high tide line. Like the other tropical islands, Soldado is heavily vegetated, bounded by a mangrove forest on its landward side (fig. 6.8).

All kinds of evidence indicate that Isla Soldado is very young—much younger than the neighboring Isla Santa Barbara. Despite the rapid growth rate of the forests there, none of the trees are large. In fact, the average diameter of tree trunks is only six to eight inches. The one exception is a small grove of large black mangroves, and their presence tells another story of the island. The black mangroves probably reached their height when they grew in the lagoon *behind* the island. When the island migrated landward, the large mangrove trees came to the ocean.

FIGURE 6.6

A ditch excavated to determine the geologic history of the island. The reeds surrounding the ditch are among the most salt-tolerant plants on the island and are restricted to a band along the open-ocean shoreline.

FIGURE 6.7 *(above)*
Aerial view of Soldado Island, an extraordinarily young barrier island. The transgressive and rapidly subsiding island is mostly less than fifty yards wide and has a very recent vegetative cover, devoid of large trees. Judging from the age of buried garbage that floated out from Buenaventura Bay, it may be that the exposed part of this island that is above normal high tide level is less than fifty years old.

FIGURE 6.8
The mangrove forest behind Soldado Island. Sediment-trapping by the mangroves builds up the elevation of the lagoon floor and is an important part of barrier island evolution. To the right is a termite mound. *Photo by Bill Neal.*

A few spadefuls of sand from the island reveal that despite its dense vegetative growth and a surface layer of leaves and plant detritus, there is almost no soil development—no dark, loamy humus layer at all. In this tropical climate, soils may develop over the sand in just a few decades—but there is no such soil here. Isla Santa Barbara, on the other hand, has a well-developed soil profile.

FIGURE 6.9
The hotel on Soldado Island in Soldado Village at the island's south end. Erosion forced the abandonment of the hotel a year before this picture was taken in 1995. The high tide beach can be seen in the background behind the hotel. A banner welcoming guests still hangs on the front of the building.

Aerial photographs tell even more. A comparison of photos taken in 1961 and in the 1990s show startling changes. For example, the southern third of the island formed entirely in only thirty-five years or so. The shoreline in the 1990s is now *behind* the lagoon shoreline of the 1960s, testimony that the island's southern third has migrated a distance equal to its entire width in that short span of time. At the same time, the northern end of the island has accreted slightly. Sand eroded from the rapidly eroding south end is pushed to the north by longshore currents.

At the south end, on the outskirts of Soldado village, some ramshackle wooden buildings on stilts fell victim to the erosion and now resided on the beach. One of the buildings had been a hotel for a single year before it was undermined (fig. 6.9). A new-appearing white cloth banner with red lettering welcoming guests was still on the building. It was an ill-fated attempt to bring tourists to a barrier island. We learned that plans were currently afoot to build cottages on Isla Soldado for wealthy citizens of nearby Buenaventura, in spite of this obvious evidence of island instability.

The most unusual evidence indicating Soldado's age as thirty to forty years is buried garbage. It is unfortunate for the local environment but fortunate for our studies that the town of Buenaventura empties much of its solid waste directly into Buenaventura Bay. When the tide and winds are just right, masses of floating plastic, cans, and bottles stream into the sea, littering the beaches. When we excavated trenches across the island to examine the layers of sand, we found items of garbage as deep as three feet below the elevation of the high tide swash line on the beach. We assumed that the trash was buried by the wave-borne overwash sand that created the island's elevation, but we deduced that the remnant garbage was probably post-1960s vintage, based on the presence of such items as Styrofoam cups and bottles from local brands of beer.

Incredible as it still seems, all the available evidence told us that Isla Soldado is made up of sand that was deposited only since the 1960s.

It is important to note that while the ocean beaches near the mouth of Buenaventura Bay are unspeakably littered, most of Colombia's Pacific coast beaches are clean, pristine, and beautiful.

We believed that certainly some of the sand making up both Isla Soldado and Isla Santa Barbara had been deposited by tsunamis. One of the major goals of our study, supported by the National Geographic Society, was to find out if such events were recorded in the sediment column and if so, to determine the frequency of tsunamis in the past. We hoped that we would be able to tell the local people the rate at which they might expect the killer waves to occur. Although we tried hard, especially in our trench-side conferences, we did not succeed. We never could identify tsunami deposits with any certainty. The sand here is very uniform in size and color and contains few shells. Uniform sand provides nothing to segregate and produces poorly defined layers at best. Another possibility is that the roots of the tropical vegetation that took hold immediately after a tsunami erased the evidence of the natural catastrophe.

Later, when we sledgehammered cores into the island, we were able to determine that the thickness of the overwash sand was usually at least six feet. Below that sand, cores penetrated black mangrove mud, full of roots and wood fragments. The cores told the story. Soldado is a transgressive island (described in chapter 2 and shown in fig. 2.3). It is migrating toward the mainland, over the mangrove forest, which explains the black mangrove mud below the overwash sand deposits.

Isla Soldado is rapidly sinking—or the sea level is rapidly rising—which seems to be the fate of all barrier islands on this stretch of coast. Rough calculations indicate that the sea level at Soldado has been rising at a rate of ten feet per century over the last few decades. This is much faster than the rising sea levels of one foot per century along the Atlantic Coast of North America, or the four feet per century on the Mississippi Delta.

As the Colombian barrier islands sink, they are buried and preserved in place. This is extremely unusual for barrier islands. For example, the Louisiana barrier islands end up as a dispersed sand layer on the continental shelf. The combination of rapid sinking and huge sand supply ends up preserving the Colombian islands beneath the waves.

Will Isla Soldado survive? It may be the most dynamic island in a chain of sixty-five very dynamic islands. These fragile strips of land are subject to deep-seated earth forces that diminish the islands as they sink and build the mountains as they are pushed upward. Working in opposition are the weathering and erosion forces on the surface of the earth that constantly lower the mountains and furnish new mica-rich sand to build the islands. The islands either shrink or grow, depending on whether sand supply or island sinking is dominant. They are never stable, but they are extremely durable.

The vast geologic energy of this coastal region makes the Colombian barrier islands unique in the world.

The Carbonate Islands
Tropical Permafrost

THE three island chains unveiled in this chapter would, at first glance, seem to have little in common. They are the islands along the northern tip of the Yucatán Peninsula, Mexico, the desert shore of Abu Dhabi in the Persian Gulf, and the Bazaruto Archipelago of Mozambique, East Africa. The common thread that binds them is calcium carbonate. This chemical compound, precipitated out of seawater, either inorganically or by marine organisms, plays a number of roles in the evolution of these islands. The grains that make up Yucatán and Abu Dhabi Islands are entirely calcareous. Bazaruto sand is mostly quartz grains rather than calcium carbonate, but the beaches and dunes of Bazaruto and the Abu Dhabi islands are extensively cemented by calcium carbonate that precipitates out of seawater. It is their carbonate connection that makes these islands unique and fascinating.

THE YUCATÁN: FLAMINGOS AND JAGUARS

In the early 1950s Chicxulub, a small Mayan village near the north coast of the Yucatán Peninsula, was chosen by the Pemex oil company as the site to drill an exploratory well to find out why the company's instruments registered a strange gravity anomaly in the area. After a few hundred feet of drilling, rock that was broken up, fused, and even melted, appeared in the core samples. Although the true and startling significance of this discovery wouldn't be known for decades, one Mexican geologist speculated that they had discovered some kind of meteorite impact crater.

In 1980 a father-son geologist team, Luis and Walter Alvarez, proposed that sixty-five million years ago a giant asteroid had struck Earth, producing a dust cloud of huge proportions that changed climates and killed off more than half of the animal species on Earth's surface, including the dinosaurs. It was a daring suggestion, based as it was on a half-inch-thick layer of clay in an outcrop in Italy. The layer was right at the boundary between the Cretaceous period and the younger Tertiary period, the point in geologic time that had already been established as the date of the dinosaurs' disappearance. The question was, Why did the dinosaurs disappear at this time? The Alvarez team gave a possible answer. Now the question became, Where did the asteroid hit?

Fifty years after the Pemex hole was drilled, the Chicxulub crater, as it has become known, proved to be the smoking gun. A planetary body, perhaps twelve miles in diameter, crashed into the shallow ocean of what is now the southwestern Gulf of Mexico. A crater 110 to 180 miles wide was created, and geologic dating shows that it was formed sixty-five million years ago. The explosive impact on what is now the Yucatán Peninsula was huge, releasing a great mass of vaporized rock that circled the globe and triggered a series of events, including global cooling, that killed vulnerable creatures in the seas as well as on the land.

In the ensuing sixty-five million years, more than three thousand feet of limestone rock has been deposited, from the sea, on top of the crater, but the faults and fractures that control the groundwater system of the peninsula are still governed by the crater shape. For example, sinkholes (ceynotes) on the peninsula, depressions of ten to hundreds of yards across that were caused by the collapse of caves in limestone, are found lined up in arcs corresponding to the rim of the buried crater. Even the location and shape of the Yucatán Peninsula and its shoreline were partly determined by this great catastrophe of sixty-five million years ago.

The thick layer of limestone deposited since the asteroid impact sent pieces of the Yucatán flying all over the world is now the Yucatán coastal plain. It is entirely calcium carbonate. Rainwater falling on this calcareous terrain is largely absorbed by the rock, through the network of caves, cavities, and fissures. The entire region is drained by the flow of groundwater to the sea, which gives rise to a number of freshwater springs welling up from the ocean floor, seaward of the barrier island chain. When limestone weathers, it tends to dissolve rather than breaking up to form sand grains like granite. So for all practical purposes there is no sand to be carried to the sea, no streams to carry the sand to the sea, and no valleys for the

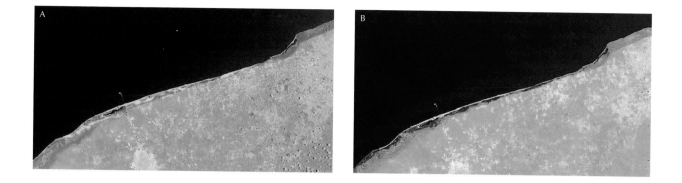

streams to run in. All the water is underground, much of it drained through the arcs of sinkholes along the rim of the Chicxulub crater.

One can hardly imagine worse soil conditions than these. The soils are thin and very rocky and provide the raw material for thousands of miles of limestone-boulder walls that function as fences, often skillfully constructed without the use of cement. Soils are patchy and restricted to scattered depressions in the limestone. Why did the great Mayan Empire flourish in such infertile conditions? The Mayan descendants, who form the largest part of the coastal population of the Yucatán, today eke out their living from the soil with great difficulty.

Limestone is responsible for the unusual nature of the Yucatán barrier island system. The islands form a continuous narrow line along the coast, behind which is a discontinuous but always narrow lagoon. The location of Progreso, the largest city (population, 45,000) on the northern Yucatán barrier island chain, coincides with the widest lagoon, about one mile, which has allowed extensive dredging to form a small port. Lagoon width is quite variable because lagoons expand considerably during the wet season (figs. 7.1a and 7.1b) and after storms and floods. Tide range is less than three feet here, but wind probably causes more changes in water levels in the lagoons than tides do.

Barely a quartz grain can be found on the barrier islands stretching for 250 miles along the northern and northwestern tip of the Yucatán Peninsula (fig. 7.2). Beach and island sediment is entirely made up of shell fragments, pushed ashore by the waves coming in from the continental shelf (fig. 7.3). Lagoon sediment is also entirely calcium carbonate but is made up of a mixture of shell fragments from the open ocean and very fine, often invisible to the naked eye, crystals of calcium carbonate, precipitated by algae.

Because of the absence of river valleys, the Holocene sea level rise did not form a deeply embayed, crooked shoreline (as it did on the U.S. east coast)

FIGURE 7.1, A and B
Two satellite images of the northwestern corner of the Yucatán Peninsula. One (A) from the dry season (July) shows a small lagoon. The other (B) photo taken during the season of high rainfall (November) shows a much more extensive lagoon. The light-colored urban area south of the shoreline is the city of Merida, Mexico. A long pier extends seaward from Progreso, the largest city on the barrier island chain. The peninsula is entirely limestone, rainwater drains into caves and cavities, and as a consequence there are no river valleys for rising sea level to flood and form an irregular shoreline. Hence the barrier islands of the Yucatán closely hug the mainland shoreline. *Landsat 7 images from the U.S. Geological Survey.*

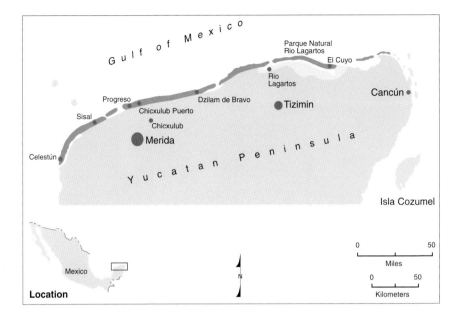

FIGURE 7.2 *(above)*
Map of the northern tip of the Yucatán Peninsula, Mexico.

FIGURE 7.3
Storm coming ashore on a typical Yucatán barrier island beach. The sediment is light-colored because it is 100 percent calcium carbonate, made up of fragments of calcareous marine organisms such as mollusks, barnacles, coral, algae, and sea urchins. *Photo by Dave Bush.*

for the barrier islands to straighten out. Instead, the shoreline of the flooded coastal plain had only the most minor of irregularities, probably caused by eroded sinkholes bordering it. As a result, the shoreline-straightening process of barrier island formation produced a barrier island chain hugging the mainland. The islands almost certainly originated as spits, growing from east to west.

FIGURE 7.4
The lagoon behind the islands of the Yucatán Peninsula is largely a mudflat occupied by water only during the wet season and after spring tides and storm surges. The mud has an orange and sometimes red tint because of the incorporation of brine shrimp and other organic matter in the sediment.

Instead of open water, the "lagoons" are mudflats that are frequently inundated by salt water. Often the mudflat surfaces are a spectacular yellow-reddish color formed by an organic mat of algae and diatoms combined with dead brine shrimp (fig. 7.4). Large areas of the lagoons, especially in the Rio Lagartos Natural Park, are used in commercial salt production. Diked evaporation lagoons within the natural lagoons are constructed in the time-honored fashion of marine salt production around the world. Even in the natural parts of the lagoon, exchange of water with the sea is limited, and evaporation can make the water as much as five times saltier than seawater.

The hypersaline, warm, and shallow water conditions bring thousands of flamingos to nest behind the barrier islands. They have become a major Yucatán tourist attraction. It's not always an easy life for flamingos in spite of their strict protection by Mexican authorities. Besides hurricanes, the birds have to worry about jaguars. On a recent visit to Rio Lagartos Park headquarters, I observed the feeding by hand of some fifty baby flamingos, all hatched from abandoned eggs rescued by park rangers. Each female lays a single egg, but the nesting area for an entire flock was abandoned because of a jaguar attack. The park superintendent, while distressed by the attack on one flamingo flock, said he was pleased to learn that they had jaguars. And the baby birds looked well fed.

The islands in this low-wave-energy, low-tidal-amplitude environment formed as spits built in an east-to-west direction. Most are eroding now, but in the past some island widening has occurred, as evidenced by beach ridges in island interiors that once lined the beach. The Yucatán island chain has both transgressive and regressive islands.

Seawalls are burgeoning in front of Yucatán beach communities, but strong opposition to the walls is growing from those concerned (correctly) that the walls will wipe out sea turtle nesting sites. The islands range between twenty-five yards and half a mile in width and are generally less than fifteen feet in elevation.

No natural inlets remain. At least three artificial inlets have been excavated to provide access to marinas in the lagoon. Two others have been dredged to the center of the island where new marinas for the commercial fishers were gouged out of the island. Jetties, all causing very rapid erosion on the adjacent shorelines to the west, have stabilized all five of these inlets (fig. 7.5).

It is likely that without the human presence, the barrier bars would be true islands only after storms. Between storms the volume of water in the lagoon is probably too small to maintain inlets.

Island vegetation is very dense and difficult to walk through. Roads, sometimes narrow, bumpy, and high-centered, are found on all northern Yucatán Peninsula islands. Once while driving a particularly narrow road in one of the parks, we ran into a black-uniformed four-man patrol of the Mexican navy. They questioned us rather closely before waving us on, probably still suspicious about our motives for looking at remote parts of the islands. They all carried rifles and looked like they meant business—except that they were accompanied by four dogs that seemed to be the standard

FIGURE 7.5
Offset of shorelines on a Yucatán jetty. All jetties along the Yucatán shoreline indicate a strong east to west longshore current, causing sand to pile up on the updrift east wall as the shoreline erodes on the downdrift west side. *Photo by John Clark.*

THE CARBONATE ISLANDS

small, short-haired, nondescript, very quiet and shy variety that abound in the Yucatán. I would have expected magnificent police dogs to go with the officers' magnificent uniforms.

More than a decade ago another major natural event occurred on the Yucatán Peninsula in the Chicxulub region. Although less energy—by many orders of magnitude—was released by this incident, it certainly was of major importance to the barrier islands and the people of the Yucatán. Hurricane Gilbert, a storm of very large diameter, roared ashore near Cancún, the peninsula's east coast resort city, on September 14, 1988. As a category 5 hurricane with a record low atmospheric pressure of 26.15 inches, it was one of the mightiest storms in history. The storm crossed the peninsula and exited again into the Gulf of Mexico, with the eye passing not too far from Chicxulub. In nearby Progreso and in other communities, the first row of buildings was sometimes virtually smashed apart. Because the construction was largely of concrete, not easily moved even by storm waves, rubble from the destroyed first row piled up into a more or less continuous "seawall" that protected the second and third rows of houses.

The maximum storm surge was eighteen feet. Where dense natural maritime forest remained in place, relatively little overwash occurred. But in developed areas, devoid of forest, overwash was frequent and damaging to buildings and roads. When the storm surge ebbed or returned to the sea, it formed spectacular surging rivers, especially where it was hemmed in between buildings or walls or in dune notches (fig. 7.6). The standing waves

FIGURE 7.6
An excavation in a blacktop road near Progreso, Mexico, caused by the passage of Hurricane Gilbert. The scooping-out of the pavement occurred during the storm surge ebb, the return of storm-driven waters back to the sea. The erosive force of the flowing waters was increased by its confinement between walls and buildings, which acted like the sides of canyons. *Photo by Dave Bush.*

formed on these surging, mountain stream–like seaward flows actually excavated regularly spaced troughs across shore-perpendicular paved roads.

The barrier islands reacted as most barrier islands would. Where the islands were narrow and low, at least twenty-six new inlets were established (fig. 7.7), mostly at the eastern end of the island chain. None of the inlets remain. Eventually, they probably would all have closed naturally, but the salt company immediately closed many of them by bulldozer and dump-trucked sand (fig. 7.8).

It was a miracle that only twenty-nine people died in a class 5 hurricane, although two hundred thousand were made homeless. Mexican officials did a good job of evacuating citizens from the barrier islands, in some cases using armed soldiers. One New York tourist couple was strolling on the Cancún beach, enjoying the solitude and the strange weather at the same time that the rapid evacuation was occurring. Returning to the hotel, they walked through a curiously quiet lobby and within a few hours were huddling in terror in the bathroom of their room, the room with a great view of the sea. They survived to tell their tale.

THE ABU DHABI ISLANDS: SABKHAS AND OOLITES

Tucked in the southernmost corner of the Persian Gulf is a chain of eight or more (depending on what one counts as an island) barrier islands (fig. 7.9). They line a fifty-mile section of the Abu Dhabi coast, better known as the Trucial Coast and once known as the Pirate Coast (batik 7.1). These are the hottest, driest barrier islands in the world. On Abu Dhabi Island sits

FIGURE 7.7 *(left)*
One of the twenty-six inlets formed during Hurricane Gilbert. Most inlets filled in naturally after the storm because the flow of water to and from the lagoons is too small to maintain an opening, since the tidal amplitude is small and fresh water flow to the lagoon is negligible. Under these conditions, longshore-transported beach sand quickly fills in the newly formed inlets. *Photo by John Clark.*

FIGURE 7.8
The narrow barrier island section near Rio Lagartos Park at the east end of the barrier island chain where most of the inlets formed during Hurricane Gilbert. The open ocean is to the left. After the storm, the salt companies reconstituted this part of the island by bulldozer and dump truck and rebuilt the dikes (to the right) used in salt production in the lagoon.

THE CARBONATE ISLANDS

the city of Abu Dhabi, population 930,000, the largest city on any barrier island. It is a modern city replete with parks and gardens, wide thoroughfares, high-rises, international hotel chains, and major port facilities.

Abu Dhabi, the emirate, is the largest of the seven kingdoms that joined together in 1971 to form the United Arab Emirates, a nation the size of Maine. Surrounded by Saudi Arabia and Oman (fig. 7.10), this prosperous nation of 2.5 million people annually exports nearly $40 billion worth of petroleum.

In huge contrast to the Colombian barrier islands, with their two-hundred-plus inches of annual rainfall, these barriers have less than one inch per year. Accompanying the very low rainfall are the high temperatures of the Arabian Desert, the warm nearshore ocean waters of the Persian Gulf, and the high salinities of the Persian Gulf waters. Normal oceanic salinity is somewhere around 35–36 parts per thousand of dissolved salts, in contrast to open-gulf salinities of 42–45 parts per thousand and 55–65 parts per thousand in the lagoons behind the barrier islands. Shamals, the severe desert windstorms, come mainly from the northwest, essentially perpendicular to the barrier island shorelines. Tidal amplitude is around eight feet along the open Persian Gulf shoreline. Like the Yucatán barriers, the Abu Dhabi barrier island sand grains are all calcium carbonate.

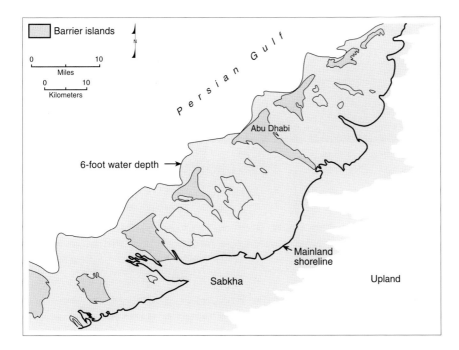

FIGURE 7.9
Map of the Abu Dhabi Persian Gulf Islands. Seaward extensions of the six-foot underwater contour line show the location of the ebb tidal deltas.

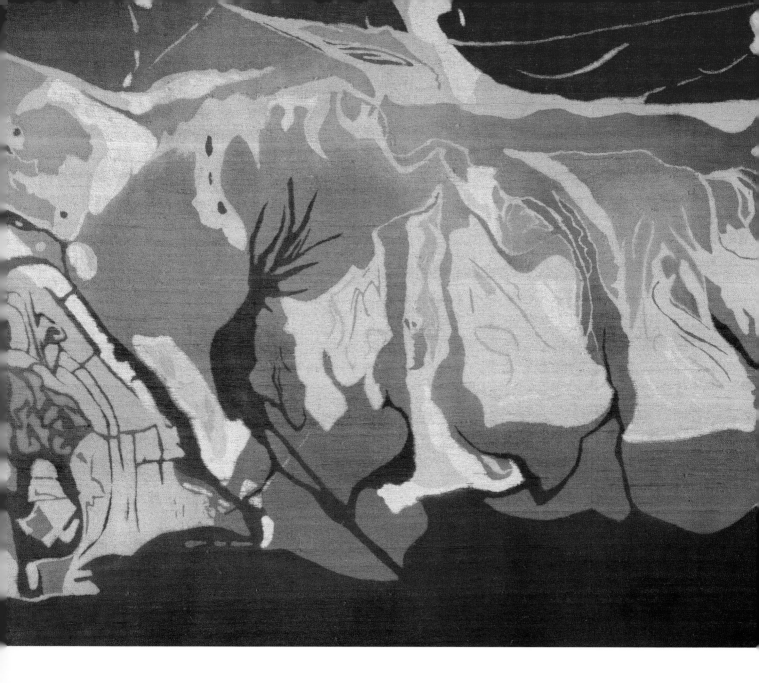

BATIK 7.1

Abu Dhabi's Arid Islands, Persian Gulf, 2001, 34" x 43"

These striking desert islands are pure calcium carbonate. Some of the sand grains are oolites, which precipitate out of seawater on the island's tidal deltas. The triangular shape is due to the accumulation of wind and storm washover materials in the lagoon, held in place by films or mats made by marine organisms. The uncharacteristic sharp angles and straight lines on the outline of some of these islands is caused by beachrock, naturally cemented beach sand.

Artist's note: The *Abu Dhabi's Arid Islands* batik comes from a satellite image found on the cover of the Society for Sedimentary Geology's Photo CD-2 of the arid region carbonate-evaporite systems. As an artist, I liked the format, the violet, turquoise, and sand colors, and the triangular islands.

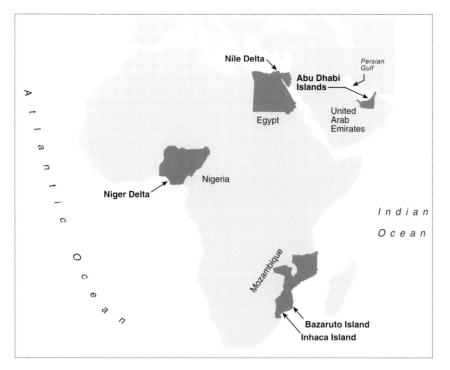

FIGURE 7.10

Map showing the location of the Abu Dhabi Islands plus the African barrier island chains discussed in this volume.

The Abu Dhabi barrier island chain formed atop a rocky offshore ridge almost parallel to the shoreline, called the Great Pearl Bank. The bank was the primary site of the pearl industry, the economic mainstay of the Emirate of Abu Dhabi before the coming of piracy and later the coming of oil. Between the Great Pearl Bank and the mainland is a low trough occupied by the Kor al Bazm Lagoon.

These islands did not originate by migration from offshore with a rising sea level, like the Yucatán barrier islands probably did. Instead they "nucleated" around ancient coral reef islands that once grew in a higher sea level, three thousand or four thousand years ago. Sand was pushed ashore from the nearby shallow waters and then carried laterally by longshore currents to lengthen the islands. Outcrops of the coral rock, originally responsible for the location of the islands, are still visible on some islands, but there are lots of other rock outcrops that are not of coral origin. This rock is called *beachrock*, and it plays an important role in island evolution.

Beachrock forms a long, narrow band, tens of feet wide and one to three feet thick, down the length of a beach between the high and low tide lines. The rock is beach sand, cemented in place by calcium carbonate precipitated out of seawater. It has the same layering that is in the beach, the same sand components. Anything in the beach, including bottle caps and glass fragments, may be cemented in the rock. I once found a crescent wrench cemented in beachrock along the North Coast of Puerto Rico. These tropical and subtropical phenomena occur where the water is clear and warm and the surf zone or current activity is vigorous.

When a band of beachrock forms along the beach of a barrier island, the processes of island evolution instantly change. Beachrock is to tropical barrier islands what permafrost is to Arctic barrier islands: the beach is hardened. Suddenly there is a natural seawall or jetty, retarding erosion and island or inlet migration. The storm response of the beach changes, the amount of sand blown into the dunes changes, and the amount of longshore transport of sand changes. It's a whole new island-evolution ball game.

The extensive beachrock outcrops on the Abu Dhabi islands are responsible for the blocky shapes, the sharp corners, and the straight lines in the island outlines (fig. 7.11). Normally, the outlines of barrier islands are smooth and rounded. In several instances in the Abu Dhabi islands, beachrock that originally formed along the margins of an inlet now extends offshore like a jetty, left there by the retreat of the ocean shoreline, which exposed the beachrock that once lined the inlet.

The triangle shape of the Abu Dhabi barrier islands is unique. The bases of the triangles, which face toward the sea, consist of the dunes (some as high as thirty-five feet) and overwash fans typical of barrier islands every-

FIGURE 7.11

Satellite image showing almost the entire Abu Dhabi barrier island chain. The dark band, bordering the mainland shoreline, is a sabkha, a low, flat zone, flooded occasionally by spring tides and storm surges. The city of Abu Dhabi can be distinguished by the dark pattern on the island to the right of center. Note the straight lines, sharp corners, and narrow lines that extend seaward of the shoreline on some of the islands. This is beachrock, which after formation has been left behind as shorelines retreat or adjust in various ways. Some of the extensions of beachrock act as natural jetties or groins, affecting the flow of sand on beaches and through inlets. They also hold the inlets firmly in place. *Shuttle photo courtesy of Gene Shinn and the Society for Sedimentary Geology.*

THE CARBONATE ISLANDS

where. The landward-pointing tip of the triangles is a mud and sand flat extending into the lagoon from the island. The flat was formed by wind-blown sand taken from the island by the dominant northwestern winds as well as by mud-and-sand-bearing floods during storms and spring tides. Most of the sand-size particles of these sediment tails are shell fragments. Probably much of the mud is derived from the dust storms so famous in this region, and the rest is carbonate mud precipitated by algae and other organisms. The sediment tails forming the triangles behind the islands are slowly expanding, so eventually the shallow lagoons behind the Abu Dhabi islands will fill in (as will all lagoons everywhere if sea level will hold still).

Why the triangle shapes? Why don't the lagoonal tidal currents sweep the island-derived sediment away and redistribute it, as they do in back of other barrier island chains around the world? The answer is not clear. Probably it has to do with organic mats, films, or "slimes," formed atop the sediment. The mats, made of algae or diatoms, lightly cement the sand in place, to be moved only by storms.

On the mainland, a nine-mile-wide *sabkha* borders the shoreline. Sabkhas are flat, featureless surfaces without vegetation that occasionally flood by storm surges and spring tides. When saltwater flooding occurs in this hot, dry climate, minerals precipitated out of seawater (evaporites) form.

Among the major constituents of the sediment of the Abu Dhabi islands are oolites, almost perfectly spherical sand grains of calcium carbonate (aragonite), precipitated out of seawater. The ebb tidal delta shoals, which continually feed the beaches and adjacent islands, are giant production sites for this sand (fig. 7.12). Oolites are responsible for the spectacular bright white color of the ebb tidal deltas.

Such grains are found in warm, shallow, and wave-agitated waters in many areas of the world. Most famous perhaps are the oolite deposits on top of the Bahama Banks, which are being considered as a source of beach nourishment sand for the barrier island beaches of east Florida. The Abu Dhabi barrier islands are among the few with the capability of manufacturing their own sand.

MOZAMBIQUE: CROCODILES AND MONKEYS

It wasn't a total surprise to see two young men with AK-47s walking past our tourist cottage on Bazaruto Island. After all, this was Mozambique, a country that finished a terrible civil war ten years back and whose people were still

FIGURE 7.12
A closer satellite view of the Abu Dhabi Islands centered on large Ras al Khaf Island. Channels and ebb tidal deltas are clearly outlined in this photo. Oolites, spherical sand grains precipitated out of seawater and made of calcium carbonate, form on the wave-agitated ebb tidal shoals. The oolites are then incorporated into the islands. In effect, these islands are continually manufacturing a part of their sand supply. *Landsat TM image courtesy of Gene Shinn and the Society for Sedimentary Geology.*

suffering the consequences of unmapped minefields. I was however, very disappointed. I was not looking forward to wandering along isolated stretches of this remote island when it was infested with armed young men.

First impressions can be very wrong. Unbeknownst to us, our arrival coincided with the sinking and breakup of a large fishing vessel. The ship had collided with an offshore ridge of beachrock that perhaps a thousand years ago had defined a shoreline position of Bazaruto Island, half a mile seaward of the present beach. The vessel was loaded to the gunwales with marijuana. The two armed men were policemen, a rather modest official response to an event that would have attracted a small army of enforcers along any North American or European shoreline.

The marijuana was quickly recovered by divers and carried off somewhere on a police barge, to be burned, we were told (fig. 7.13). The Pakistani crew was captured after they rowed fifteen miles to the mainland. Wooden fragments of the vessel washed up on the beach and were quickly recycled by the local citizens. Within days the island returned to its usual peaceful state, and we never saw another AK-47.

Mozambique (fig. 7.9) has a 1,550-mile shoreline, about one-third of the entire East African seaboard. Although it was a Portuguese colony until 1975, its proximity to South Africa has resulted in a strong Anglophile flavor, including driving on the left side of the road and membership in the British Commonwealth.

FIGURE 7.13
Offshore ridges formed by beachrock left behind by the retreating shoreline (see line of waves in fig. 7.17) are a menace for the unwary mariner. During our study of Bazaruto, a marijuana-laden ship ran aground on such a ridge. The ship's cargo was salvaged and is shown here in the process of being loaded on a police barge.

THE CARBONATE ISLANDS

Ilha da Bazaruto and Ilha da Inhaca are two extraordinary Indian Ocean barrier islands in Southern Mozambique, lying between Madagascar and Mozambique, within five to twenty miles of the Mozambique mainland. The islands are in the Indian Ocean southeast trade wind system, which generates large waves and strong south to north longshore currents. In addition, a narrow and steep continental shelf minimizes wave friction with the seafloor and adds to the wave height.

Inhaca (In-yak-a) Island, the better-known (geologically) of the two, is seven miles long and five miles wide and is situated eighteen miles offshore of Maputo, the capital of Mozambique, close to the border with South Africa (fig. 7.14). This unusual width-to-length ratio is due to the fact that Inhaca probably is a composite of two Ice Age barrier islands that are currently being blanketed by Holocene sands, blown in from the beach. Bazaruto Island, 325 miles to the north of Inhaca, is eighteen miles long and up to three miles wide. The "Bazaruto Archipelago" consists of three barrier islands (batik 7.2, fig. 7.15), two of which, Magarugue and Benguerua,

FIGURE 7.14 *(left)*
Map of Inhaca Island, Mozambique.

FIGURE 7.15
Map of the Bazaruto barrier island archipelago, Mozambique.

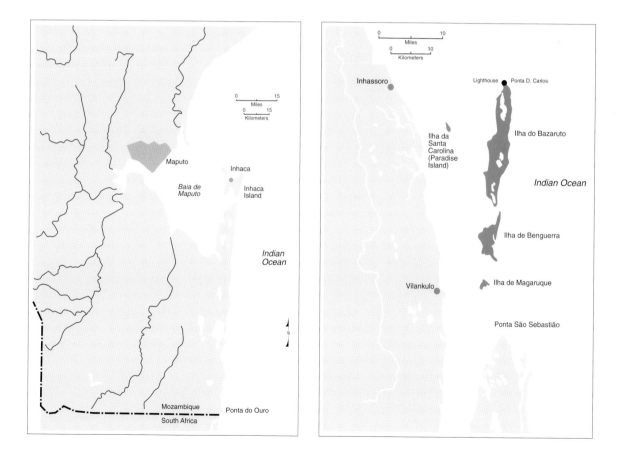

BATIK 7.2

Bazaruto's Giant Dunes, Mozambique, Africa, 2001, 30″ x 36″

Most of the volume of sand that makes up this island accumulated during the Ice Age. Beachrock has played a big role in the evolution of this eighteen-mile-long East African island. The white sand is the modern or Holocene sand, which is creeping over the three-hundred- to four-hundred-foot-high Pleistocene frontal dunes, perhaps the highest on any barrier island in the world.

Artist's note: For this batik the tables were turned and Pilkey brought home the aerial photograph of Bazaruto Island, Mozambique. I dyed the gorgeous blue lake in the center, waxed the immense white dunes on the left and the waves shaping the shore in the right background. Bazaruto is a verdant island, and I listened to indigenous music while working.

are south of and much smaller than Bazaruto. Each of the three islands of the archipelago is a single Pleistocene island blanketed on the ocean side by Holocene sands, which are slowly taking over the island (fig. 7.16).

Portuguese politics has played an unusual role in preservation of the beautiful offshore islands. Antonio Salazar, the Portuguese dictator from 1932 to 1968, was not one to tolerate opposition kindly. His way of handling troublemakers was to lure them to the colonies. As a result, Mozambique, Angola, and the other colonies gained some of the best and brightest geologists and biologists. The unintended result was an environmental sensitivity and concern for future generations that was unusual in colonial administrations. This situation led to the recognition of Inhaca Island as a natural preserve well before World War II.

After gaining independence, the country went through a horrendous civil war, in spite of the fact that the Mozambique's boundaries, drawn in 1891, had actually taken into account tribal boundaries and regional history. The war between the Communists (RENAMO) and the anti-Communists

FIGURE 7.16
Satellite image of eighteen-mile-long Bazaruto Island, along with the much smaller Benguerra and Magaruque Islands. The islands probably formed as storms breached a spit that had formed by north-moving sand from the mainland. The ebb and flood tidal deltas are quite large and have complex channel systems. Most of the irregularities in the shorelines are beachrock-related. The dominant winds from the southeast have produced the barely visible, southeast-to-northwest-trending "stripes" of alternating sand and vegetation on the ocean side of Bazaruto. *NASA shuttle photo number STS 68-275-69.*

175

(FRELIMO), which ended in 1992, resulted in the deaths of a million people. At the peak of the war, thousands of refugees made their way to both Inhaca and Bazaruto Islands, where they found safety in isolation but lived under terrible conditions, made all the worse by a very limited supply of fresh water. The islands also suffered from vast overgrazing by goats and unsustainable cooking-wood harvest. Now the islands are back to their natural-preserve status.

In a Third World country that has just finished a devastating war, natural preserves don't mean the same thing as a natural park in Europe and America. Both islands have indigenous populations living a largely subsistence existence, with all that that entails, with regard to agriculture and the need for cooking wood. Inhaca has six thousand inhabitants, while the larger Bazaruto has only two thousand. On Inhaca, a high percentage of the buildings are at least partially constructed with concrete block and tin roofs. On Bazaruto, all homes are thatched huts.

Both Inhaca and the Bazaruto Archipelago Islands formed originally as spits extending northward from mainland capes. The frontal dune ridges of both islands range in elevation from 200 to 450 feet, the largest I have seen on any barrier islands. The dune completely thwarts overwash. The Australian barrier islands have thousand-foot-high dunes, but these are composite dune fields, unrelated to barrier islands, made up of superimposed layers of dune sands of widely differing ages.

On Bazaruto and Inhaca Islands, the bulk of the sand in the beachfront dunes is Pleistocene. Holocene sand is simply draped over the 120,000-year-old dune and does not build a separate and distinct dune ridge. On Bazaruto Island the Holocene sand has marched over the frontal dune (fig. 7.17) and is beginning to move across the island toward the lagoon, gradually covering the scrub forest and filling some of the ponds.

The Holocene and Pleistocene sands can easily be distinguished by their color. The older Ice Age sand has a reddish hue imparted by a stain of iron oxide (hematite) on the mineral grains. The iron was obtained by dissolving iron-bearing heavy minerals within the sediment. The modern sediment is a striking white color with a tinge of yellow and contains shell fragments. In addition to color differences, there is often a lack of stratification or layering in the Pleistocene sands. This is caused by more than a hundred thousand years of groundwater movement through the sand that has disrupted grains and dissolved away all of the shell material (fig. 7.18). The white Holocene sand retains the distinctive cross-stratification that is characteristic of wind deposits.

FIGURE 7.17 *(above)*
Light-colored Holocene dune sand spilling over the high Pleistocene dune ridge on Bazaruto and beginning to move across the island, burying forests and ponds in the process. On the ocean side, the indentation on the beach marks where the shoreline has "overstepped" a line of beachrock. The location of the former beachrock is marked by the line of breaking waves. Notice the narrow barrier island on the lagoon side of the larger barrier island, a most unusual occurrence.

FIGURE 7.18
An eroding cliff of Pleistocene dune sand on the lagoon side of Inhaca Island. The red color, from iron staining of quartz grains, the lack of tiny shell fragments that are always present in modern barrier island dune sands, and the lack of characteristic dune layering all indicate that the sand is old, perhaps more than 100,000 years. The dune is believed to have formed on the oldest of two Pleistocene barrier islands that make up most of the volume of Inhaca Island.

All the Mozambique islands are lined by rock at the shoreline, most of it beachrock, but there is a second type of rock here that is also characteristic of tropical to subtropical shores—*aeolianite,* or cemented dune sand. Beachrock and aeolianite are easy to distinguish on the basis of grain size (dune sand is usually much finer than beach sand) and the characteristic layering of wind- versus water-deposited materials. The extent of aeolianite on Bazaruto and Inhaca is still to be determined, but clearly portions of the giant frontal dune are cemented, in some instances, thirty or forty feet above sea level.

The rock, no matter what its origin, greatly slows down the rate of shoreline retreat and stabilizes the island. Nonetheless, the island slowly retreats, leaving behind the beachrock as an offshore ridge or a kind of natural breakwater. Off Bazaruto, breaking waves mark lines of former beachrock. Frequently the stranded offshore beachrock forms the foundation for small coral reefs.

On both Inhaca and Bazaruto Islands there is evidence of a higher Holocene sea level than at present, perhaps four to seven feet higher. At South Point, the extreme south end of Bazaruto Island, the jettylike rock outcrop (fig. 7.19) includes a layer of beachrock above the high tide line formed earlier in the Holocene. It is an extraordinary and indisputable piece of evidence that the sea level was higher in the recent past. Incontrovertible evidence of higher sea level is hard to come by on most barrier islands made of unconsolidated sand.

Both Inhaca and Bazaruto are undergoing erosion at rates of a foot or two per year. On Inhaca Island, the entire ocean-facing dune front is eroding at a rate that increases to the north. Corresponding to this increased erosion is a steepening of the large dune face to the point that slumps or landslides are occurring down the face of the giant frontal dune. Every time a slump brings new sand to the beach, it is quickly removed by the waves.

Bazaruto Island has more than a dozen circular freshwater ponds, some more than half a mile in diameter. The ponds occupy low areas on the is-

FIGURE 7.19
Beachrock lining the Bazaruto side of the inlet between Bazaruto and Benguerra Islands. This extraordinary natural outcrop is as straight and true as the finest engineered jetty, and like the human-made equivalent, it holds the inlet in place and affects the sand supply on adjacent islands. Two layers of beachrock are exposed here, one at the top where the person is standing and the other at the base of the outcrop at the waterline. Between the two layers of beachrock is a layer of aeolianite—cemented dune sand. The upper layer of beachrock is proof positive of a higher sea level in the past, probably two or three thousand years ago.

land that probably originated as blowouts, formed by winds at a time when the water table was lower than it is today. Similar blowout ponds exist on many barrier islands around the world, but what makes these unique is the crocodile inhabitants. Some are as long as sixteen feet and are easily visible by day, sunning themselves at the surface. At night, two bright-red buttons in the beam of a light mark their location.

The other inhabitants of Bazaruto, rarely found on barrier islands, are monkeys. Only twenty green monkeys remain on the island, and we were fortunate to spot five of them running full tilt across a dune, tails held high. Among other things, the encounter explained some mystifying footprints in the sand.

Preservation of Bazaruto Island has become an international priority. The World Wildlife Fund has a lodge on the island with a small permanent staff. Protection of the sea turtles, reduction of the loss of forest on the island, and preservation of the dudong are the fund's three main priorities. The Mozambique equivalent of the Florida manatee, the dudong was much treasured for its food value, so much so that it has almost disappeared. The region was famous for dudong meat, and when important politicians visited a town, a dudong feast would invariably be prepared to celebrate the event.

THE CALCAREOUS ONES

With the carbonate islands, we have added a chemical dimension to barrier island evolution. Calcium carbonate plays both a chemical and a physical role in each of the island chains in Mozambique, Abu Dhabi, and Mexico—chemical because it acts like cement, precipitated out of seawater to form beachrock and aeolianite, and physical because the grains act just like quartz grains, to be sorted by wind and waves and moved in proportion to their size, shape, and weight.

All three island chains are coastal plain islands, but each formed in a different fashion. The Yucatán islands are classical coastal plain types that straighten out a slightly crooked shoreline. For the Abu Dhabi barrier islands, an old coral reef on top of a ridge (the Great Pearl Banks) formed a nucleus that accumulated sand driven ashore by waves, to eventually become barrier islands. The Mozambique islands also formed atop a subtle ridge that provided the platform for a spit to be formed by longshore currents that flow from south to north. The Yucatán islands migrated to their present location. The Mozambique and Abu Dhabi islands more or less formed in place, taking advantage of preexisting topography.

On the northern tip of the Yucatán Peninsula, the sand making up the islands consists entirely of calcium carbonate grains, mostly fragments of the shells of various marine organisms. This composition has occurred not because productivity of seashells is particularly high in Yucatán nearshore waters but because of the complete lack of dilution by riverborne quartz grains from the mainland. Beachrock, although occasionally present on the Yucatán islands, does not seem to play a critical role in island evolution.

Sand on the Abu Dhabi coastal plain islands is also 100 percent calcium carbonate, a lot of it manufactured on-site, as oolites are precipitated out of sea water on the ebb tidal delta shoals. Beachrock plays a very important role in island evolution here, as the cemented layers of sand act as seawalls and jetties. Even island shape is affected by beachrock. The islands' blocky outlines, with sharp corners and straight lines, are quite unlike the smoothly rounded contours of unconsolidated barrier islands.

To a lesser extent, beachrock also affects the island outlines of the Mozambique Islands, especially adjacent to some of the inlets. Both in Mozambique and in Abu Dhabi, beachrock that once lined an inlet now extends seaward beyond the beach as the shoreline has retreated. The former inlet-armoring rock has now become a natural jetty.

The Mozambique sand is not highly calcareous; it contains mostly between 10 percent and 20 percent calcium carbonate. But cementation of sand to form both beachrock and aeolianite has had a strong stabilizing effect on the islands. It is possible that beachrock is the explanation for the extraordinarily high dunes. Perhaps holding the island in place by cemented natural seawalls has provided time for these big beachfront dunes to form.

In the short run, so long as there is no cementation, barrier islands made up entirely of calcium carbonate work exactly like those made up entirely of quartz grains. The carbonate grains are washed across the island by overwash, are blown into sand dunes, and are carried by longshore currents along the beach. However, calcium carbonate (calcareous) grains differ from quartz grains in two geologically important respects. Quartz is very hard and is little changed by rolling around in the surf zone. Calcium carbonate is much softer, and the grains tend to become rounded quickly and are often broken up.

Second, calcium carbonate, once buried, can be readily dissolved in groundwater, but quartz cannot. As sea level goes down, stranding barrier islands on the mainland, the shell material in the sandy sediment is usually dissolved and removed. This is why the Pleistocene sediment in the Mozambique islands and the Pleistocene lagoon islands of the east coast of North America have few seashell fragments in them. When the sea level comes back up, bringing along with it a new barrier island, the new island sand is recharged with new carbonate shell fragments. But quartz grains remain and persist.

It is entirely possible that quartz grains in a Temperate Zone barrier island have been in several barrier islands previously. But carbonate grains are good for one island only.

Lagoon Barriers

The Quiet Ones

IN 1585 Sir Walter Raleigh, poet, entrepreneur, statesman, and favorite of Queen Elizabeth, sent 108 men to Roanoke Island in what is now the state of North Carolina. The Spanish, buttressed by the greatest war fleet of the time, were creeping up the east coast of North America, exploring and claiming lands, and the English were determined to stop them. Roanoke Island lies within Roanoke Sound, which is connected to the larger Pamlico and Albermarle Sounds, behind the Outer Banks of North Carolina. It was the first English settlement in North America, and it was meant as an outpost to limit Spanish intrusions. It was short-lived. The settlers, preoccupied by the fruitless search for gold and clashes with the natives, were not up to the task of forming a new settlement and were carried back to England the next year by Sir Francis Drake.

Not one to be deterred by adversity and financial loss, Raleigh sent 117 people—89 men, 17 women, and 11 children—back to Roanoke Island in 1587. The courageous settlers were each to be given five hundred acres of land in exchange for their efforts to form the "cittie of Ralegh." Sir Walter's grants of acreage to the settlers began the long European tradition of settling and claiming land already inhabited by native peoples. The settlement could not have been started at a worse time. Tree ring studies today show that the worst drought in eight hundred years occurred in the years 1587 to 1589.

The expedition was under the command of John White, who would later gain fame as an artist who depicted the ways of the natives of the New World. White stayed with the settlers for a month, long enough to celebrate the birth of his granddaughter, Virginia Dare, and to witness the conversion to Christianity of Native American chief Manteo, who had just returned from a brief visit to England, where he met the queen.

Actually, the colonists were supposed to be put off on the beach in Chesapeake Bay, but ship captain Simon Ferdinando, impatient to prey on Spanish vessels as a privateer, sailed through an inlet on the Outer Banks barrier island chain and put the colonists ashore early. Geophysical studies by East Carolina University geologist Stan Riggs suggested initially that a large inlet, open at the time, connected the sound to the ocean within a mile of Roanoke Island. This seemed like the logical place for Ferdinando's ship to enter the quiet sound waters (at the site of the present-day highway bridge from Manteo across the sound). Later studies by Riggs proved that the channel was actually an old riverbed and not an inlet. It is more likely that the ship passed through an inlet much further south into Pamlico Sound, out of sight of Roanoke Island.

John White returned to England for more supplies and more colonists, promising to return in six months, but the war with Spain prevented his return until three years later. When he arrived there was no sign of the settlers. Only the word *Croatan* carved in a tree gave a clue to their fate—a clue whose meaning is still debated today. Did Spanish marauders or the local natives kill the settlers? Did they wander off to a new settlement? Or were they simply assimilated into a local tribe? White was not able to institute even a brief search for the lost colonists because, once again, his ship captain was insistent on heading back to sea.

The Lost Colony was a failed colony on a failed barrier island. Roanoke Island, once a true barrier island, no longer faced the open ocean (fig. 1.3). The roar of the surf zone was gone. By the time the settlers arrived, the transport of sand had become a one-way system. The occasional storm took sand away from the island but never returned it. What was missing in the sound was fair-weather waves that ocean-facing barrier islands depend on, waves large enough to push the lost sand back to the beach. Roanoke Island was disappearing—and it still is today. It can't migrate; it can't recover from storms. It can only remain in place and be slowly consumed by the rising sea level.

Geologist Riggs believes that, at its north end, which faces the greatest fetch (distance of open water), more than a mile of Roanoke Island has disappeared since the Lost Colony vanished. The original site of the Lost Colony may well be underwater. In the shallow water of Roanoke Sound that surrounds the diminishing island, the seafloor is littered with a mixture of English and Native American pottery fragments, stone tools and arrowheads, pieces of early brick, and other evidences of human occupation (including beer bottles). Storms have scattered the detritus far and wide and have mingled together the relics of precolonial, colonial, and present-day occupation.

Because of its disconnect with the open-sea processes that originally formed it, Roanoke Island is a relict lagoon barrier island. It probably formed during a high sea level period, during the last interglacial, 120,000 years ago. It is important at this point to reiterate that the 120,000-year date of the last interglacial sea level high is a number still in doubt and still in debate. Among other possibilities, the last sea level high may have been 80,000 years ago. Probably a majority of coastal geologists (for whatever a majority vote is worth in science) currently accept the 120,000 number.

Many islands like Roanoke formed during the last interglacial time when the sea level was similar to the present level. A few islands are currently active, evolving in response to present-day waves and winds. But most are accidents of geologic time. They are bodies of sand that were once barrier islands, now isolated again as individual islands because the level of the sea happens to be where it is at this moment.

Even though there are many more lagoon barriers than their open-ocean cousins, they have not been studied much by geologists, perhaps because they are no longer dynamic, spectacular, and changing. The excitement is gone. The forces that formed them and maintained them have left, leaving behind only the forces that destroy them. Listed below are the types of lagoonal barrier islands. A living island is one that continues to evolve, either moving back or widening, depending upon its supply of sand and the behavior of sea level. A dead island, like Roanoke Island, is a pile of sand that has no way to maintain itself.

- **active (live) islands**, lagoon islands that form and evolve in equilibrium with present-day wind, tide, and wave conditions; identical to their oceanic siblings except that they face open-lagoon waters
- **relict (dead) islands**, former barrier islands now stranded in quiet water, including the following:
- **stranded islands**, islands formed at a previous sea level high, then left stranded when the sea level dropped and now flooded by a newly risen sea
- **cheniers**, former barrier islands, now surrounded by wetlands or marshes because of a shoreline advance caused by a surge in muddy sediment from a nearby river (Mississippi Delta) or construction of a new barrier island to the seaward (Mekong Delta)

- **inlet islands**, islands formed as waves wash through an inlet mouth, piling up sand on salt marsh or mangrove banks. If the inlet migrates, the inlet island receives no sand and becomes a dead island
- **human–made (dead) islands**, piles of sand left behind by the dredging of navigation channels, shell refuse from seafood plants, or large rocks from sailing ship ballast

ACTIVE LAGOON BARRIER ISLANDS: THE LIVING ONES

Not everyone agrees that the island known today as Roanoke Island was the Roanoke Island of old, the site of the Lost Colony. A minority historical opinion holds that what was called Roanoke Island in 1587 may actually have been what is today called Cedar Island, fifty miles to the south in Pamlico Sound. If this is the case, Roanoke Island detractors believe, the National Park Service should move its Lost Colony museum and other facilities to Cedar Island.

One argument is that Ferdinando's ship would most likely have sailed through the easier-to-navigate Ocracoke Inlet south of Cape Hatteras and closer to Cedar Island (fig. 1.3). The inlets through the Outer Banks north of Cape Hatteras would have had much higher waves and would have been difficult to navigate without charts. Ocracoke Inlet definitely existed in Ferdinando's time, because it occupies and has been constrained by an old river channel. Other evidence used to support the claim to Cedar Island's historical role, such as local folklore and pottery shards, is a bit less than convincing. But who knows?

Cedar Island is one of the best examples of a living lagoon barrier island in North America (batik 8.1). Seven miles long and less than a hundred yards wide, the island has a beach that faces northeast, the direction of maximum fetch (fifty miles) and the direction from which the biggest winter storms come. Cedar Island is the only active lagoon barrier island in Pamlico Sound, and it exists here for two reasons. First is the presence of the large fetch needed to generate waves of sufficient size to mold sand into a barrier island. Second is the availability of a large supply of sand, obtained by cannibalizing an Ice Age barrier island as Cedar Island migrates into it. This old island, with a slightly different orientation and a much larger volume of sand, probably formed facing the open ocean 120,000 years ago, at the same time Roanoke Island formed. The Ice Age island is shown in the

BATIK 8.1

Cedar Island, N.C., 1995, 69" x 36"
Cedar Island, an unusually well-developed lagoon barrier island, owes its existence to fifty miles of open water fetch to the northeast, the direction from which most winter storms arrive. Probably Cedar Island's main source of sand is the Pleistocene island behind it (shown in red-brown) that is being cannibalized as the island retreats.

Artist's note: Cedar Island was my first attempt to prove to Pilkey that batiks could be valuable illustrations of barrier islands. I asked him which of his personal favorite islands was difficult to portray. He gave me a black-and-white photograph of Cedar Island, where the ferry takes off to the Outer Banks. Charts and explanations of the strange shape of the unique island became part of the art.

batik as the darker island behind Cedar Island. The modern island obtains sand as it slowly overruns the ancient island.

Cedar Island most likely originated well to the seaward (or soundward) and has slowly migrated back to its present location. Proof of island migration was found in our core samples, taken a few years ago, which revealed that the island is a thin (three to ten feet) layer of sand on top of old compacted salt-marsh mud.

One problem with solving the mystery of the Lost Colony is that over the four hundred years that have passed since it disappeared, Roanoke Island and Cedar Island have both changed dramatically. But the ways in which they have changed provide a good illustration of the difference between dead and living lagoon islands. Roanoke Island is a dead island, rapidly and permanently losing land area. Cedar Island has changed its location by migration over past centuries but probably not its size. In either case—island destruction or island migration—the answer to one of America's great history mysteries may be underwater.

A lot of people pass through Cedar Island, since it provides the landing for the ferry going to Ocracoke Island on the North Carolina Outer Banks. A few small beach cottages and a motel adorn the shoreline near Cedar Island's center, next to the ferry landing. Most of the island is quite pristine, for reasons suggested by its nickname. Local youths refer to it as Skeeter Island because the abundance of mosquitoes during the summer is downright spectacular. Cedar Island, we discovered, is best studied during the mild winters of coastal North Carolina.

One of my research trips to the island unfortunately coincided with high summer temperatures and dead-calm winds. My graduate student Tsung Yi Lin and I thought we were well prepared. Long-sleeved shirts, long pants, gloves, a jacket thick enough to bend mosquito probes, and liberal application of insect repellent on remaining exposed areas should have provided excellent protection. Alas, the repellent seemed to be of little help, and the occasional mosquito, with a case-hardened stinger, penetrated our clothes. When finally we both began to inhale mosquitoes from the insect clouds around our heads, we retreated to the gentle surf zone and walked in the waist-deep water, fully clothed.

On this particular expedition we were checking out experimental sand traps made from PVC pipe driven into the sand. The traps were designed to capture and measure the sand that was blown about in order to help us understand the role of wind in shaping barrier islands. As we approached the traps, arms flailing through the black clouds of mosquitoes, we were assailed by a strong rotting odor. Beaches abound with dead plants and animals. The resulting organic matter plays an important part in the island's ecosystem. But in this case the odor did not come from a decaying fish, turtle, or porpoise carcass. It came from our sand traps.

It had not occurred to us that ghost crabs would tumble into the traps and be unable to escape, and other geologists who had used similar traps on beaches had failed to mention this problem. Back to the drawing board.

Walking the beach at Cedar Island, we could spot small patches of well-compacted mud at the water's edge. The mud was originally deposited in the quiet waters on the back of the island, only to be exposed as the island migrated over it. The mud patches are covered with clumps of healthy salt-marsh grass. While such patches are also very common on the beaches of migrating islands along the open ocean, they never support a flourishing salt marsh. Salt-marsh grass (*Spartina alterniflora*) can withstand occasional storms, but it cannot coexist with waves on a daily basis.

Cedar Island also offers a valuable lesson about the impact of grazing animals on barrier islands. It illustrates the huge importance of vegetation in the control of the evolutionary processes of barrier islands. Cedar Island is actually a beautiful, if accidental, experiment on the effect of uncontrolled grazing on islands, a problem of considerable interest to barrier island managers and residents around the world.

What better place to graze cattle than on a barrier island? No need to maintain fences. It would take a determined and well-equipped rustler to make off with the herd. In a temperate climate, it is easy to dig small ponds to furnish fresh water stored just below the surface, derived from rainfall.

On large numbers of barrier islands, however, escaped or abandoned livestock have multiplied to the point that they strongly affect the vegetative cover. Examples include Mustang Island in Texas, Saint Vincent Island in Florida, Shackleford Banks in North Carolina, Assateaque-Chincoteaque Islands in Virginia, and Fire Island in New York. Since natural predators have long since been eradicated from the islands, the grazing animals, which may include some combination of deer, horses, cows, sheep, goats, and pigs, reproduce to the limits allowed by the vegetative cover.

The road leading to the Ocracoke ferry divides Cedar Island into grazed and ungrazed areas. West of the ferry road, horses and cows have not grazed for perhaps fifty years or more. East of the ferry road, free-ranging horses and cows have grazed for decades, probably for centuries. The horses are wild, but the cows are said to be owned by local residents, who occasionally round them up. Among these cows range several imposing bulls, ready to challenge the unwary intruder.

Healthy, well-vegetated dunes five to seven feet high and a dense maritime forest characterize the ungrazed region, west of the ferry road. The forest, restricted to a band on the back of the island, is dominated by pines and is dense enough to make walking difficult. The ungrazed land has a green, lush appearance broken only by patches of strikingly white sand in the low areas between dunes (fig. 8.1).

In contrast, the grazed eastern part of Cedar Island lacks dunes altogether (fig. 8.2). Patches of mud are found on the beach, but no salt marsh grows on them. The maritime forest is largely gone. A few live oaks remain, starkly beautiful in their isolation. No young trees or bushes exist to impede walking among the widely spaced trees (fig. 8.3).

The environment of the grazed portion of the island is clearly very different from that of the ungrazed portion. This leads to different processes of island evolution. Both ends of Cedar Island are slowly migrating back,

FIGURE 8.1 *(above)*
Horses and cows have grazed heavily on the eastern half of Cedar Island, but the western half, shown here, separated from the rest of the island by a highway and ferry landing, has remained essentially untouched by grazers. As a result, extensive sand dunes have developed, with a healthy grass cover, along with a dense maritime forest bordering the lagoon side of the island. Comparing this scene with that of the overgrazed section of island is testimony to the importance of vegetation in island development.

FIGURE 8.2
A flock of Canada geese feed on the overgrazed eastern half of Cedar Island, where there are no dunes because the dune grasses necessary for their construction and maintenance have been entirely consumed by cows and horses. The maritime forest consists of widely spaced mature trees with no intervening small trees or bushes, which have all been removed by grazing. Removal of predators from island ecosystems has led to overgrazing on many islands around the world, not only by cows and horses, as in the case of Cedar Island, but also by goats, sheep, deer, and pigs.

190

FIGURE 8.3
Aerial view of the highly overgrazed east end of Cedar Island. The island is white because of the lack of vegetation. A narrow rim of maritime forest lines the back side of the island next to the salt marsh. Three offshore sandbars can be seen in the photo. *Photo by Jordan Pilkey.*

and the shoreline retreat rates on either side of the ferry road are identical. But each is evolving in a different way. Along the ungrazed portion, Cedar Island slowly widens as small dunes spill into the salt marsh behind. The eastern part of Cedar Island, on the other hand, is frequently overwashed, and storm waves form lobes of sand that extend across the island.

There is a plus side to the overgrazing of Cedar Island. The vast flat and duneless area provides nesting places for hundreds, perhaps thousands, of seabirds. The main danger to the nests comes from the grazing activity of nest-oblivious cows. People are less of a threat to the nesting birds—the mosquitoes have seen to that.

There are other active barrier islands in large bays around the world, usually not so well developed as Cedar Island. Small active barrier islands exist in both Delaware (fig. 8.4) and Chesapeake Bays.

RELICT LAGOON BARRIER ISLANDS: THE DEAD ONES

Forty miles to the south of Cedar Island is another lagoon barrier island, Harkers Island, North Carolina. South facing and three miles long, the island has a fetch of only two miles across Core Sound, a body of water that is mostly less than three feet deep. This lagoon exists behind Core Banks, the southernmost island of the Outer Banks.

FIGURE 8.4

Small active lagoonal barrier islands in Delaware Bay along the Delaware shoreline. The islands have migrated landward over salt marsh mud, which now resides on the beach. Mosquito ditches dug behind the islands in the 1930s are now visible in the marsh mud layer in front of the island. The ditches in front line up perfectly with the ditches behind the island, a spectacular proof of island migration in a coastal environment with low waves. *Photo by John C. Kraft.*

Harkers Island clearly did not form under present-day conditions. Waves large enough to move sand and construct a large island could not form in such a shallow, narrow body of water (fig. 8.5). In fact, Harkers Island is an ancient body of sand formed 120,000 years ago. It was once part of the same chain as the Ice Age island now cannibalized for sand as Cedar Island migrates into it. The fact that Harkers Island is completely surrounded by water is purely an accident of the current sea level, just as in the case of relict Roanoke Island.

Harkers Island is also an interesting study in isolated living. As a seagoing scientist in the 1960s and 1970s, I relished the rich Elizabethan English, Outer Banks accent of the native residents who were crew members on the Duke University vessel RV *Eastward*. The islanders have a long seafaring history and tradition and often provide crew for research vessels, dredges, tugs, and Coast Guard ships.

The well-to-do shore-loving crowd did not discover the island until twenty years ago. Now the homes of the wealthy sprout along its seawalled lagoon shoreline, and the local accent is heard with decreasing frequency.

Nearby Beaufort and Morehead City, North Carolina, occupy former islands from the same ancient barrier island chain as Harkers Island. The early settlers chose to build on these old islands because they provided elevation and relative safety from storms. The barrier island origin explains why these towns are stretched out in a long, narrow line.

There are numerous other ancient inactive lagoon barriers along the U.S. East and Gulf Coasts and along the coasts of Mexico and Brazil. South Carolina has a particularly large number of lagoon islands (batiks 8.2, 8.3, 8.4). One of them, Long Island, is situated behind Folly Beach, South Carolina. Nine-mile-long Saint Helena Island, located behind Fripp Island, South Carolina, is the largest dead lagoon barrier island in the eastern United States. Skidaway Island, behind Wassaw Island, Georgia, a five-mile-long, two-mile-wide barrier, is the home of the Skidaway Institute of Oceanography. No longer do waves, even small ones, pound the shorelines of these islands (fig. 8.6). Long Island, Skidaway Island, and Saint Helena Island all are now almost entirely surrounded by salt marsh rather than open water.

FIGURE 8.5 *(above)*
The "oceanfront" of Harkers Island, N.C. This lagoon island facing Core Sound is an Ice Age island behind Core Banks. A few lagoon islands are as heavily developed as this one is, but characteristically, the beach is missing, replaced by seawalls, as is the case here.

FIGURE 8.6
An island in the marsh behind little Tybee Island, Georgia. This Holocene island was apparently an open-ocean barrier island briefly, before a new island chain was built further seaward. *Photo by Jim Henry.*

BATIK 8.2 (above, 2 panels)
Kiawah at Twilight, S.C., 1995, 72" x 78"
Former beach ridges, now covered by
forest and surrounded by marsh, form
striking green patterns in the salt marshes.

Artist's note: Kiawah at Twilight was
commissioned by residents who live on
this South Carolina barrier island just
south of Charleston. My aerial photograph
reveals pine forest fingers stretching into
the marsh. This island is only a few
minutes flight time from the local airport
near my abode. The sunlight, coloring the
wetlands, is a creative addition.

BATIK 8.3
North of Johns Island, S.C., 1995, 55" x 49"
This batik nicely illustrates the classic
drainage patterns of salt marshes and shows
two lagoon barrier islands with different
orientations and probably different origins.
Perhaps one formed in the lagoon and the
other was once an ocean-facing island.

Artist's note: North of Johns Island is a
scene near my home on James Island Creek.
Vegetables used to be delivered to
Charleston via my tributary until the Corps
of Engineers closed it to boat traffic. You
can feel the twisting of the tidal creeks
in this constantly changing environment.
I wanted the island to feel like it was floating
like a lily pad in the marsh.

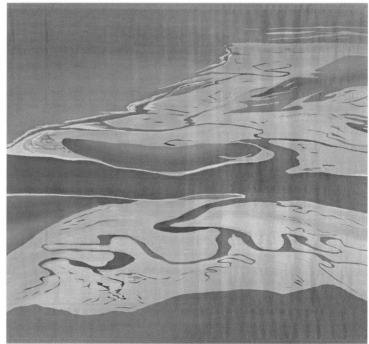

BATIK 8.4
Second Sister Creek, S.C., 1996, 84″ x 54″
The forested lagoon barrier in the center of
the batik is a stranded island that once
faced the ocean, possibly formed in the
Holocene, although the age is not certain.
It is completely surrounded by salt marsh.
The remarkably preserved recurved spits,
the curved extensions of the islands, were
formed as the island grew (from top to
bottom in the batik) by spit extension.

Artist's note: Second Sister Creek has
morning sun reddening the banks of the
sandy island in the foreground. This pine
forest on an ancient dune ridge is behind
Folly Beach, South Carolina. Marine
biologist Dorsey Worthy piloted a 172
Cessna, and my trusty FM2 Nikon caught
this aerial landscape.

In a larger sense, the relict islands in lagoons are just one component of a vast network of stranded barrier islands left on the coastal plain from times of higher sea level (fig. 8.7). On geologic maps, the ancient islands are referred to as scarps because early geologists who mapped these features did not recognize them as landlocked ancient barrier islands. They were usually named after a nearby geographic feature or town, such as the Trail Ridge Scarp in Florida, the Orangeburg Scarp in South Carolina, and the Suffolk Scarp in North Carolina. The islands now in the lagoons behind the real barrier islands are simply those that have been flooded by the present sea level. If sea level continues to rise and the barrier islands migrate landward up the rim of the coastal plain, other stranded barrier islands will emerge in the newly formed lagoons.

Cheniers

First identified in the Mississippi Delta in Louisiana, cheniers are named for the oak trees that blanket them. Called *chene* by the Cajuns, they are sand ridges encased within a salt marsh or other kind of wetland. Cheniers are former islands that have been abandoned by the sea as a result of deposition of river sediment on their seaward side. So much river sediment arrives in front of the island that the waves don't have time or the energy to remove it, and a mudflat (covered by salt marsh or mangroves) is formed. If the supply of river sediment is once again interrupted, perhaps as discharge channels on the active delta lobes change position, waves will attack the marsh-covered mudflat, erode the shoreline, remove the mud, and concentrate sand. The mud drifts into deep water while the sand stays behind to be shaped into a new island. The original island, now behind the new island, has become a chenier.

Good examples of this phenomenon are the cheniers at Cameron, Louisiana. Here a series of more than a dozen sand ridges, separated by thin bands of marsh mud, mark the three-thousand-year history of shoreline change that is related to variations in the sediment supply from the mighty Mississippi. It is likely that cheniers exist on many other deltas of the world as well, including those of the Gurupi River and the Mekong River (fig. 5.11).

Cheniers can be a few hundred yards wide and a few miles long. They have proved to be fertile farming zones but are precarious places upon which to live. Since they are often less than ten feet high, they have a strong tendency to flood during hurricanes and typhoons.

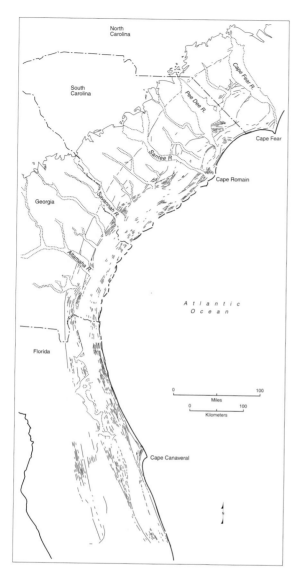

FIGURE 8.7

A map of former barrier island chains left stranded on the coastal plain of the southeastern United States over the last million years or more. As the glaciers come and go in the high latitudes, sea level falls and rises. Each time the sea level reaches a maximum and begins to diminish, some kind of evidence of a shoreline, usually a barrier island, is left stranded on the coastal plain. The complex series of lines on this drawing represent possible former barrier islands. *From Winkler and Howard (1977).*

Inlet Islands and Maoris

Inlet islands are a special kind of lagoon barrier island that forms immediately inside inlets, where ocean waves roll into the lagoon. Batiks 8.5 and 8.6 show the Otter Islands, which are inlet islands inside the entrance to Saint Helena Sound, South Carolina. Batik 8.7 depicts inlet islands in Bulls Bay, South Carolina. An extraordinary example of this island type is found on New Zealand's North Island, not far from Auckland. Inland from the mouth of Kaipara Harbor, an inlet between two Pleistocene (Ice

BATIK 8.5

South Edisto River, S.C., 2000, 156" x 36"

The Edisto River meanders through large areas of salt- and freshwater marsh to finally arrive at the sea. To the right of the river mouth is the tip of Edisto Island. The islands in the center are the Otter Islands, inlet islands well within Saint Helena Sound that are similar in origin to Tapora Bank, New Zealand.

Artist's note: The Edisto River is in the Ashepoo, Combahee, and South Edisto (ACE) Basin of South Carolina, one of the largest undeveloped estuaries on the east coast of the United States. A nautical chart from the National Ocean Service was employed for reference. I gave this design a giant Japanese woodblock print resemblance. The barrier islands depicted are Otter Island, Pine Island, and Edisto Island.

Age) barrier spits, is Tapora Bank (fig. 8.8), at the landward rim of the flood tidal delta. This island is two and a half miles long and half a mile wide, with two parallel sand dune ridges fifteen to sixty feet high, all formed within the last hundred years.

In the 1960s Tapora Bank was surrounded by water at all tide levels (the average tidal amplitude here is seven feet). Today Tapora Bank is surrounded by water only at high tide, because windblown sand from the island has slowly filled in this lagoon within a lagoon. The mainland behind the island is constructed of a series of Tapora-like islands, each of which successively attached itself to the mainland as its lagoon became dry land.

New Zealand coastal scientists have learned to take quite seriously the legends of the native Maoris, Pacific Islanders who arrived on the island a thousand years ago. In this case the Maori legends explain the anomalous occurrence of a layer of marsh mud on the beach, which shouldn't be there, since the island has never migrated. The recent geologic history of Tapora Bank is well understood from historical photos and charts, and the usual explanation that mud on the beach originated as the island migrated over marsh mud doesn't work. Tapora Bank clearly is not a transgressive island (fig. 2.3). New Zealand geologists Terry Hume and Quentin Smith believe that the Maori legend that long ago an island existed in front of Tapora Bank would explain the Tapora beach mud layer. The old marsh layer was behind the island of the Maori legend. The island may have been destroyed by a big storm that left behind bits and pieces of marsh mud that now appear on the Tapora Bank beaches.

FIGURE 8.8
New Zealand's Tapora Bank formed on the edge of the flood tidal delta inside Kaipara Inlet. New Zealand coastal scientists have used Maori legends to help explain the geological history of this lagoonal island. The island is less than a hundred years old. *Photo by Quentin Smith.*

Back in the United States, Bill Cleary, coastal geologist of the University of North Carolina at Wilmington, has identified another type of lagoon inlet island, examples of which are found behind Wrightsville Beach. It is a very small-scale version of Tapora Bank island. These are forested ridges of sand, a few tens to hundreds of yards long and perhaps ten yards wide. They are usually enclosed on all sides by salt marsh (fig. 8.9).

Initially these little islands were assumed to be dredge spoil islands from the distant historic past. I was so certain of this that I wagered a bottle of red wine with Professor Cleary, who won the bet when his research indicated that natural processes formed the islands during a time when an inlet existed just to the seaward.

As big ocean waves rolled through the throat of the inlet and into the lagoon, sand was pushed atop the eroding salt marsh to form a small island. As the inlet migrated, waves would continue to pour through and push more sand on the salt-marsh banks, lengthening the original small island over the distance of inlet migration. When the inlet closed, as inlets frequently do, a long, thin ridge of sand remained in the lagoon with enough elevation to allow vegetation to flourish. Most likely, inlet islands are just a few hundred years old. The process of their formation continues today at Masons Inlet at the north end of Wrightsville Beach.

BATIK 8.6 *(opposite)*
Edisto IV, S.C., 1999, 64" x 40"
Another perspective of the Otter Islands with Saint Helena Sound in the background. These islands, shown as dark, forested strips along the edges of the marsh, form with the energy and sand provided by the open-ocean waves that enter the sound.

Artist's note: An excursion over the ACE Basin south of Charleston offered numerous aerial photographs of land meeting water. My brother, Burke, and I were flying in the Ercoupe, banking Edisto Island at my wingtip. The resulting batik indicates a constant shift of the sands in response to waves and tide.

FIGURE 8.9
A thin, narrow marsh island in the salt marsh behind Wrightsville Beach, North Carolina. The marsh island formed when an inlet was at this location and waves pushed sand on top of the marsh. When the inlet migrated away, the island remained in the marsh. The lagoon shoreline of Wrightsville Beach can be seen at the bottom of the photo. The Intracoastal Waterway along the mainland shore at the top of the photo is lined with small dredge spoil islands, marked by dark green vegetation. *Photo by Bill Cleary.*

People can't seem to help themselves. Lagoons are easily accessible bodies of water, individuals don't own them, and when you throw something into the water, it sinks and will never be seen again. In Colombia and perhaps many other parts of the world, large volumes of solid waste are disposed of in lagoons. Out of sight, out of mind. All over the coastal world, humans dig, dredge, or fill in lagoons behind barrier islands to make navigation

BATIK 8.7
Bull Island, S.C., 1985, 60" x 45"
In the background at the top of the batik is Bull Island, a "hot dog" island that faces the open ocean. In the foreground are several narrow and low islands that are in Bulls Bay, subjected to much lower waves.

Artist's note: Before Hurricane Hugo the placid landscape of Bull Island was just gorgeous from the bird's-eye view. I flew back shortly after the 1989 storm surge and found that winds had wreaked havoc with the shoreline and the maritime forest. The trees were broken like matchsticks, tossed randomly about, and the beauty took years to restore.

channels, to dispose of shells from harvested seafood, to build causeways for highways, or to acquire sand for construction and beach nourishment. In the old days, sailing ships often sailed "in ballast" and upon arrival dumped the ballast rocks, forming small islands in the lagoon (fig. 8.10).

Until the mid-twentieth century and the growth of the environmental movement, dredging was a hit-and-miss operation. Anyone could hire a local dredge to dig a channel and pile up the sand anywhere.

Dredging of channels in salt marshes for mosquito control has produced thousands of tiny islands of material, known as dredge spoil, often with a small cluster of bushes on them. Construction of canals through the marshes of Louisiana has produced hundreds of miles of tiny long and narrow islands bordering the individual navigation channels.

FIGURE 8.10

Islands made of sailing ship ballast along what is now the Intracoastal Waterway behind Sapelo Island, Georgia. The islands were created in the early nineteenth century as captains of sailing freighters that came from Europe dumped their ballast in the salt marsh. They then sailed home with lumber. Fish camps with wooden walkways and docks are now located atop the old ballast piles, which consist of rocks traceable to English quarries. *Photo by Jim Henry.*

Along the U.S. East and Gulf Coasts, the unwanted dredged material from navigation channels, often too muddy to be used for concrete or beach nourishment, is piled on "spoil islands." One purpose of island construction is to reduce the amount of sediment that gets redistributed by waves and currents and carried back into the dredged channel. Not counting the tiny islands made by dredging mosquito-control channels, more than four hundred dredge spoil islands sit in North Carolina lagoons behind the state's twenty-one barrier islands.

Human-made islands in lagoons are not barrier islands or former barrier islands by anyone's definition. The problem is that without careful study, some are virtually indistinguishable from natural islands; this is especially true once the islands become vegetated.

Some human-made islands serve as increasingly important wildlife refuges. Small dredge spoil islands tend to be free of humans, dogs, cats, and other predators that interfere with bird nesting. As a consequence, the islands have evolved into important bird sanctuaries, made all the more critical by the accelerating loss of alternative nesting land due to coastal development.

THE QUIET ONES

Live and actively migrating lagoon islands such as Cedar Island, North Carolina, and the Delaware Bay islands of Delaware form a distinct minority among the quiet ones. Most lagoon islands that are not of human origin are dead former barrier islands that once faced the sea during some past sea level high. The appropriate conditions for barrier island formation have occurred repeatedly during the ice ages, but most of the former islands were left stranded further up on the coastal plain, rather than in lagoons. Many, such as Roanoke Island, North Carolina, and the islands in Laguna Madre, Texas, are eroding away in an irreversible process of land loss, but others, such as the lagoon barriers of Georgia, are surrounded by salt marshes that slow the land loss. In all cases, these former barrier islands are not growing, not migrating, not widening, not replacing lost sand. A simple count of all islands behind barrier islands would reveal many more made by humans than made by nature, and the production of such islands, mostly as convenient places to dump dredge spoil from navigation channel maintenance, continues today, only slightly abated by environmental regulations.

The Icelandic Islands
Of Fire and Ice

WHEN Wagner wrote his *Ring* operas, based in part on the Icelandic sagas, he wasn't thinking of sandur islands, but he could have been. The dramatic music of the ride of the Valkyries could well accompany the formation of the sandur islands of Iceland as volcanoes burst through glaciers, volcanic cones fill up with meltwater and then catastrophically break, and roaring cascades of water rush to the sea to form the islands.

The great Skeidararsandur Jokulhlaup of 1996 began with small earthquakes just north of the Grimsvotn Volcano in southeastern Iceland. Grimsvotn, a volcano beneath the Vatnajokull Glacier, consists of a caldera, or basin-like depression, left behind after a volcano blows its top. The ice on top of the Grimsvotn Caldera was eight hundred feet thick before the eruption began at ten o'clock in the evening on September 30. The column of erupted ash and steam (fig. 9.1) that, thirty hours later, broke through cracks in the ice, extended thirty thousand feet into the atmosphere.

By October 1, several depressions or subsidence bowls had formed in a line on the glacier's surface. Their location indicated that the eruption was emanating from the same three-mile-long fissure as a previous large 1938 eruption. The fissure was north and upslope from the caldera and was covered by ice that was twelve hundred to two thousand feet thick. Eventually, the subsidence bowls collapsed to reveal a spectacular thousand-foot-deep vertical-walled ice canyon. Clearly visible at the canyon's bottom was a roaring stream of meltwater flowing downslope into the caldera. The caldera then began to fill with water, pushing up the thick layer of ice covering it.

FIGURE 9.1
An Icelandic volcano erupting beneath a glacier. Such eruptions are responsible for sudden huge releases of water and sediment, called glacial bursts, or jokulhlaups, that build out the barrier islands on the rims of sandurs. *Photo furnished by Richard S. Williams and the U.S. Geological Survey.*

Sandur is the Icelandic word used to describe a smooth apron of sediment, an alluvial fan that spreads out in front of a glacier instead of a mountain. Streams flow across the sandur, carrying the meltwater that discharges in summertime from glaciers (fig. 9.2) and flows through inlets to the sea. The Skeidararsandur lies in front of the Skeidara Glacier, an arm of the much larger Vatna Glacier, the largest of Iceland's 120 glaciers. At least ten sandurs exist along this stretch of the southern Iceland coast (fig. 9.3), the largest of which is the Skeidararsandur.

On average, a sudden and catastrophic release of a large amount of water from a glacier, called a jokulhlaup, flows across the Skeidararsandur every six years. Heated rocks on the floor of the Grimsvotn Caldera continually melt the ice to form a subglacial lake that periodically flows beneath the ice to the Skeidararsandur. The more spectacular jokulhlaups, as in 1996, happen when a volcano erupts under the glacier.

Iceland is a 490,000-square-mile piece of the seafloor that has been built up through the activities of more than two hundred volcanoes that cover the island. Sitting virtually astride the North Atlantic Mid-Ocean Ridge, Iceland is slowly growing by the spreading of the seafloor. Here new seafloor forms at the rate of about an inch a year and adds volume to the gigantic slabs or plates of rock that constitute the earth's crust. Eastern Iceland rides east on the European Plate, while western Iceland moves west atop the North American Plate. Iceland is one of the most tectonically active places

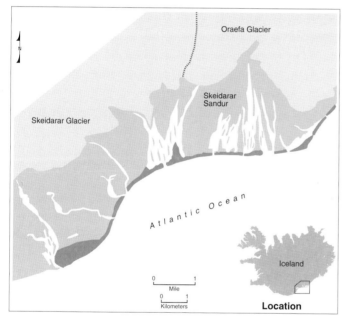

FIGURE 9.2 *(above)*

An aerial view of the Breidamerkursandur, a short distance to the west of the Skeidararsandur shown in the batik. The complete sandur system is visible. Lobes of melting glaciers can be seen in the upper part of the photo. Streams from the glaciers flow across the sandur to the small lagoons behind the islands and then through the inlets and out to sea, where the plumes of sediment-laden water are easily visible. *Photo furnished by Richard S. Williams and the U.S. Geological Survey.*

FIGURE 9.3

Map of the Skeidararsandur in the southeast corner of Iceland, showing the barrier islands that line the rim of the sandur.

on earth. Earthquakes, small and large, abound, along with volcanic eruptions, lava flows, fumaroles (steam vents), and hot springs. Like the Hawaiian Islands, Iceland sits on top of a *hot spot*, where a particularly large plume of molten rock, extending from deep within the earth, rises close to the surface of the seafloor. Iceland exists because of the enormous amount of lava spewed forth from the hot spot that happens to emerge directly below the Grimsvotn Volcano, the site of Iceland's most frequent volcanic activity.

Most of Iceland's 250,000 inhabitants closely watched the events of the October 1996 eruption. Everyone knew what had happened during an eruption at the same site in 1783. As cubic miles of basalt lava poured from fissures, huge amounts of sulfur-rich gases were released to form an "acid haze." In the next two years, 9,350 Icelanders died from the volcanic gasses, water poisoning by fluorides, and famine from crop failures.

This time, glaciologists were certain that a jokulhlaup was about to burst across the Skeidararsandur. On the first predicted date, October 4, reporters and scientists from around the world stationed themselves on the high ground next to the Skeidararsandur to watch the big event unfold. While common knowledge and common sense said the jokulhlaup would start just as the caldera filled up to the brim, that didn't happen. The melting water actually continued to rise above the lip of the caldera. The ice held the rising water in place until its level was much higher than the point at which the scientists had assumed the event would begin. In the old days there was a farmer named Ragnar who, it was said, could smell the beginning of a jokulhlaup. His was the most photographed nose in Iceland, but he was no longer around in 1996, and his daughter Anna thought she smelled one, but it was weeks before it happened.

After two weeks of failed predictions, the experts stopped making any, and some of the assembled reporters and scientists departed for home. Something unforeseen and unexpected was happening.

On October 14 the eruption ceased, although glacial ice continued to melt and meltwater continued to pour into the caldera lake. The ice cover of the Grimsvotn Caldera also continued to bulge. By October 18 the caldera lake water level was a full 160 feet higher than ever measured before.

In the morning hours of November 5, 1996, the weight of the water was so great that it literally forced its way between the lip of the caldera and the overlying ice. Suddenly it all let go, sending water rushing down toward the sea (fig. 9.4). The water carved and melted a thirty-mile-long combined tunnel and canyon under the ice in eleven hours before finally bursting out at the end of the Skeidararjokull Glacier (fig. 9.5) to suddenly and spectacularly cover the entire Skeidararsandur. The largest jokulhlaup in recorded Icelandic history was under way.

Within three days the Grimsvotn Caldera Lake was almost completely emptied; the lake level fell 550 feet in that time. At its peak, late on November 6, the flow was a twenty-mile-wide band across the Skeidararsandur, a water volume on the same order as Africa's Congo River.

When it was all over, a sandur covered by numerous multi-ton chunks of ice removed from the Skeidararjokull Glacier's snout was left behind. Some

of the ice blocks were thirty feet high and weighed more than a thousand tons. Six miles of Iceland Highway 1 disappeared, and six more were severely damaged. Two bridges were destroyed, including 225 yards of the 1,000-yard bridge across the Skeidarar River, the longest bridge in Iceland. Not a single person was killed or even injured by this colossal "natural catastrophe," a testimony to a society that listens closely to its environment. By November 7, when the jokulhlaup ceased rather abruptly, only icequakes—tremors caused by ice collapsing into the emptied caldera—remained.

The November 1996 jokulhlaup carried perhaps 100 million tons of sediment to the shoreline and beyond. A plume of discolored water was visible seven miles out to sea. The sand-laden water would finally reach the sea after it roared through the inlets of a chain of barrier islands (fig. 9.6).

THE SANDUR ISLANDS

Devoid of the trappings of civilization, lacking in vegetation and lining one of the world's highest-wave-energy coasts, the Skeidararsandur Islands of Iceland are unlike any barrier islands on the planet (batik 9.1, fig. 9.7). Formed on the rims of sandurs, they feature no tidal deltas, no tidal inlets, no sand dune beach ridges, no estuaries, no marshes, and no tidal flats. Sea level change plays no role in island evolution, and island processes are not controlled by vegetation. The islands are blessed with a huge supply of sand from nearby glaciers, which actually forces them to migrate toward the sea.

FIGURE 9.4 *(left)*
A May 27 photo, famous among Icelandic scientists, showing the Skeidararsandur Jokulhlaup of 1938. In this view to the east, the entire sandur is submerged except for the tops of the barrier islands, now broken into numerous small islands. Because the glaciers have been melting at an accelerated rate for the last few decades, the enlarged sand supply has enlarged the islands. Consequently, the 1996 jokulhlaup formed fewer new inlets, even though the flow of water was greater than that during the 1938 event. *Photo by Palmi Hannesson, furnished by Richard S. Williams and the U.S. Geological Survey.*

FIGURE 9.5
Streams emanating from Skeidara Glacier. The streams originate as single-channel flows at the nose of the glacier (upper part of photo) and then split into hundreds of branches called distributaries. In this way the sediment in the streams is spread over a large area on the sandur. *Photo by Dag Nummedal.*

FIGURE 9.6
Distributaries on the Breidamerkursandur flow directly into the sea through barrier island inlets. The surf zone shown by the white band marks the seaward boundary of the barrier islands. Some distributaries flow parallel to the lagoon shoreline before reaching the sea. Such streams erode the backbarrier shoreline, a part of the unique seaward migration process of these barrier islands. *Photo furnished by Richard S. Williams and the U.S. Geological Survey.*

The seaward rim of each of Iceland's sandurs is lined with barrier islands, totaling between fifty and seventy-five in number in any given year. With each jokulhlaup the number probably changes as new inlets form, and during the winter when beach sand transport is strong, inlets may close. Perhaps sandur barriers can be considered a special kind of delta barrier, since they are located where short glacial streams terminate and flow into the ocean. In Iceland they exist on one of the most energetic coasts in the world. Summer wave heights are commonly ten feet, but the winter waves are truly memorable. The storms that strike the south Icelandic coast,

BATIK 9.1

Iceland's Volcanic Shores, 1998, 37" x 47"

The barrier islands of Iceland are unvegetated bodies of volcanic sand derived from nearby glaciers. Sand supply is so large that the islands actually migrate in a seaward direction. The lagoon, at least during the summer, is fresh glacial meltwater.

Artist's note: Pilkey assigned the reading of many scientific papers about the sandur islands. We were going through numerous slides at Duke University when we came upon Dag Nummedal's photograph that I used for reference. The black volcanic sand is dyed a Charleston green in the batik, and we were drawn to the subtle variations in the ocean's colors.

FIGURE 9.7
Aerial view of the relatively thin and narrow barrier islands on the eastern part of the Skeidara Sandur. A small flood tidal delta can be seen, but there are no ebb tidal deltas on these islands because the extremely energetic waves prevent their formation. *Photo by Dag Nummedal.*

mostly between November and February, commonly produce forty-foot waves, breaking just offshore of the beaches.

These barrier islands are not locked in by a frozen sea during the winter, as are those in the Arctic. Nor do permafrost and floating sea ice play a significant role in barrier island evolution. Just to the south of the sandur coast, a major ocean current, a continuation of the Gulf Stream called the North Atlantic Drift, provides a strong moderating influence on climate. In Reykjavik, the capital city, 150 miles to the west, the mean January temperature is 31 degrees Fahrenheit, a surprisingly mild measure for a country located just south of the Arctic Circle. The north coast of Iceland is much colder. Occasionally, polar bears float over on ice floes from Greenland, 270 miles away, and disembark on the land there.

Two major meltwater stream systems flow across the Skeidararsandur of southern Iceland. The twenty-mile-long Skeidarar River flows to the sea to the east, and the Hverfisfljot, the Sula, and the Sandgigjukvisl Rivers combine to empty through the western barrier islands that rim this fifty-mile stretch of sandur shoreline. As the streams cross the sandur their single channels divide up into a maze of shallow, smaller stream branches that fan out across the sandur's broad stretches. When they reach the backside of the barrier islands, the stream branches turn sharply and flow along the back of the islands until several streams combine to break through and form an inlet.

Because of frequent flooding and overwash, the islands are not habitable. There are a few emergency shelters for shipwrecked mariners, equipped with food, bunks, and radio, but most shelters are located at regular intervals on the sandur plain behind the islands. On the crest of the islands, signs with arrows point the way to the nearest shelter. The main travel artery, Iceland Highway 1, follows the entire sandur coast, situated a few miles back from the shoreline. A few temporary dirt track roads cross the sandurs between the main highway and the islands.

On maps, inlets are named, but islands are not, a situation similar to the tropical barrier island chain of Colombia. Perhaps the islands are not recognized with names because they are not recognized as actual islands. During the winter, when the flow of glacial meltwater across the sandur is halted, the "lagoons" may be very shallow and sometimes are absent altogether.

The Treasure Hunt

Geologists—those outside of Iceland, at least—first learned of these islands through work done in 1973 by a group of six University of South Carolina coastal geology students and faculty. Led by Miles Hayes and Dag Nummedal and sponsored by the U.S. Naval Oceanographic Office, the expedition produced landmark studies of an unusual barrier island chain distinguished by active glacier sediment sources, extremely high waves, huge sand supplies, and no vegetation.

Actually the 1973 expedition was a treasure hunt. A group of wealthy Icelanders were eager to find the remains of a Dutch ship carrying treasure from the Dutch West Indies that went aground on the Skeidararsandur in 1667. Historical accounts indicated that the ship's masts were visible at sea for years. In fact they were not only visible, they served as an important landmark on an otherwise featureless plain. The question was whether the hull and its treasure remained in the sandur or had simply been dispersed in pieces during some past jokulhlaup.

The 1973 expedition occurred at the time of the Cod War between Great Britain and Iceland, and Iceland was threatening to pull out of NATO. The U.S. Navy, eager to please and to preserve NATO's submarine bases in Iceland, agreed to fund the treasure hunt. The intent of the University of South Carolina researchers was not to find the treasure but to determine the location of the 1667 shoreline and thus pinpoint where to concentrate the search for the wreck.

Al Hine, a graduate student at the time and now an oceanographer at the University of South Florida, remembers the expedition fondly and recalls

Iceland as a land of geologic extremes—extreme wave heights, extreme sand volumes, extreme floods. A crew of five Icelanders worked with the scientists. The expedition had at its disposal a helicopter and an ad hoc truck/bus vehicle that used wheels salvaged from B-29 bombers. An emergency station on the sandur, half a mile behind the island chain, provided housing for the whole party.

The work could be dangerous. While measuring flow rates, Hine and Nummedal were spilled into the ice-cold water of one of the river channels when their small rubber boat capsized. Both made it ashore, but barely. Once, as they were preparing to take the same boat through a rushing inlet, their Icelandic companions quietly intoned, "You won't come back if you go through the inlet." They dropped the idea.

One of the unexpected hazards was birds, particularly the Arctic skua, which Hine describes as a cross between a crow and a turkey. These birds dive-bombed people walking on the sandur, and in one case delivered a slash across the face of one of the scientists. The birds would also attack helicopters in flight, making it necessary to ascend quickly to heights greater than those frequented by the birds (and to descend with equal rapidity).

Geology of the Islands

In the early 1970s, when the University of South Carolina geologists conducted their survey, they counted a dozen Skeidararsandur islands. Rising ten to twenty feet above sea level, the islands ranged in length from three to six miles. The number and length of the islands vary considerably, depending on the recent history of river floods and the supply of sand. Jokulhlaups carve out new inlets, but the enlarged supply of sand from the retreating glacier (sixty feet per year) in recent decades apparently has widened and bulked up the islands, making it more difficult for new inlets to slash through them.

U.S. Army maps made in 1942, during the World War II occupation of Iceland by Britain and the United States, showed twenty-three islands along the rim of the Skeidararsandur. By 1951, maps showed sixteen islands. In 1977 ten islands line the shore. Both the 1942 and the 1951 maps show many more river channels crossing the sandur than today.

The number of islands also changes according to the seasons. Each winter, when water flow from the glacier is low and winter storms drive strong longshore currents to the west, sand spits at the ends of the islands extend in a westerly direction. When this happens, inlets migrate or even close. Come summer, more rivermelt water from the glaciers combines with lower ocean waves to allow the inlets to reopen in their old locations.

These could be called drumstick islands, since they are widest at their east, updrift ends and taper down at their west, downdrift ends (batik 9.1). In Iceland, bulbous island ends form on a steady, continuous basis as river waters drop their large sand loads upon arrival at sea. The sand is gradually pushed ashore, and the east end of the island, where the freshly arrived sand initially accumulates, gradually widens seaward. When the winter storms arrive, much of the summer's growth is lost to wave erosion.

The sands of Icelandic barrier islands are a striking black color because they consist of pulverized volcanic rock, mostly basalt and obsidian. No splashes of green from grasses or trees relieve the uniform ebony vistas of these vegetation-free islands. Grasses have temporarily sprouted only on inactive portions of the sandur well behind the islands (fig. 9.8). There, clumps of a long-bladed grass (*Elymus arenarius*) are common enough to allow for sheep grazing.

The inlets that separate sandur islands are, strictly speaking, not true tidal inlets, since tidal currents do not shape them. Instead, they are river mouths. The vertical difference between high and low tide level ranges from three and a half to seven feet, but even the extreme tides seldom reverse the flow of river water to the sea, as happens in the inlets of other barrier islands. Here, the fresh water flows constantly to the sea, even as the tides rise. Although the currents never reverse, the tides nonetheless cause the lagoons to expand as the river water level behind the islands rises as much as three feet.

FIGURE 9.8
Small patches of dune vegetation atop the dunes, found on temporarily abandoned portions of the sandur. *Photo by Dag Nummedal.*

Since tidal currents don't move sand out to sea, there are no tidal deltas. One might expect river deltas at the inlets, but these don't exist either, at least not for long. When a large amount of sediment reaches the shore, as during a jokulhlaup, small deltas of sand are built at the inlets, but they don't last long in the high wave environment. Deltas are quickly dispersed by the winter waves, which spread the sand along the islands and down across the shoreface. In the past, jokulhlaups have extended the shorelines of some Icelandic sandurs in a seaward direction as far as two miles. Gradually such shoreline extensions are pushed back by storm waves, but shoreline straightening may take a number of years.

The total volume of sand transported in the surf zone of these islands may be around five million cubic yards (about five hundred thousand dump truck loads) annually, by one estimate. Sometimes it is much larger. In any event, the volume of sand moved by Icelandic longshore currents is one to two orders of magnitude larger than that moved on U.S. east coast beaches. Most sand is carried to the west. The sand supply is highest on the western downdrift islands of the Skeidararsandur, and those islands are higher and wider than others.

The outermost sandur plain immediately behind the barriers is a wind tidal flat that is constantly smoothed and flattened by the seemingly never-ending winds of southeastern Iceland. Occasionally, strong winds blow water out of the river channels, flooding the surrounding flat areas. In areas not flooded by wind tides or flattened by a recent river flood or jokulhlaup, low sand dunes have formed.

The March to the Sea
The bulk of the subaerial (above sea level) portion of the sandur islands is made up of berms, sand dunes, and overwash fans. Berms are long sandbars that march out of the surf zone, up onto the beach, and eventually above the high tide line during the summer. These are features that can be seen on almost every beach in the world.

On most of the world's barrier islands, sand dune ridges parallel the shoreline. Not so in Iceland. The western islands of the Skeidararsandur have fields of small, ten-foot-high dunes that are oriented perpendicular to the shoreline (fig. 9.9). These are called transverse dunes, and when the islands are entirely overwashed, the water flows between the dune rows. No dunes of any kind form on the lower-elevation islands of the eastern part of the Skeidararsandur shoreline because these islands are overwashed too frequently.

FIGURE 9.9
Transverse dunes on a segment of one of the sandur islands. Most barrier islands around the world have ridges of dunes that parallel the shoreline. On these islands because of dominant wind directions and the high frequency of overwash, the dunes are lined up perpendicular to the shoreline. *Photo by Dag Nummedal.*

The shoreface in front of these Icelandic islands extends to water depths as great as 350 feet, perhaps the deepest and steepest barrier island shoreface in the world. The geometry of the volcanic slopes is the reason for this extraordinarily deep and steep shoreface. The 350-foot depth of the base of the shoreface contrasts with 30- to 50-foot depths typical of the U.S. East Coast islands and 6 or 7 feet for some Arctic barrier islands in Siberia.

The Skeidararsandur shoreface is long and smooth, with a few sandbars on its upper portions. It is slowly building seaward, and as it does so, the islands move with it. Fair-weather waves constantly pump sand to the islands, pushed to the beach from the upper shoreface. Winds and breaking waves bring the sand from the beaches to the islands. As Icelandic islands build out on the seaward side, they simultaneously lose sand on the other side as rivers cut into them. The net result, in huge contrast to other barrier islands, is the inexorable island migration *toward the sea*.

Over thousands of years the sandur barrier islands of Iceland must have migrated for miles, keeping pace with the gradual buildout of the sandurs. In recent centuries, the Skeidarar barrier islands along the eastern edge of the sandur have remained stationary. Rates of seaward island migration increase gradually to the west. Near the western margin of Skeidararsandur, a two-thousand-ton Norwegian trawler, originally beached in the surf zone in 1927, now stands isolated on the sandur, more than three hundred feet inland. The wreck marks the position of the 1927 shoreline.

Seaward growth of Icelandic sandurs can be downright spectacular. Fifty miles west of the Skeidararsandur, the community of Vik, on the western margin of the Myrdalssandur, had a good harbor a hundred years ago. Today the town stands three thousand feet from the sea and is harborless.

OTHER SANDURS

Sandur barrier islands are found in a few other places in the world. A few may exist in front of Alaskan glaciers that are close to the sea. It is even possible that some may exist in the Antarctic, although none have been discovered yet.

Robert Taylor, a geologist with the Canadian Geological Survey, has described sandurs on Bylot Island in the new far north Canadian province of Nunavut (fig. 9.10). Bylot Island is deep within the Arctic, much farther north than Iceland, which lies just below the Arctic Circle. Three sandur islands enclose Middle Bay, north of Cape Burney, at the mouth of an unnamed valley filled by glacier E-67, which extends to within five miles of the sea. The lack of a more exciting name for the glacier is understandable in a province that covers one-fifth the land area of Canada and has a population of only 25,000.

FIGURE 9.10
The island chain at the edge of the sandur of glacier E-67 on Bylot Island, Nunavut, Canada. The glacier that provides the sediment for the islands is clearly visible up the valley. These islands are deep within the Arctic, where, unlike southeastern Iceland, winter sea ice and permafrost play a role in their evolution. *Photo courtesy of the Canadian Geological Survey.*

THE ICELANDIC ISLANDS

Similar to the Icelandic sandurs, the Cape Burney Islands are low, frequently overwashed, unvegetated, and have small sand dunes on their upper surfaces, oriented perpendicular to the shoreline. But the Canadian islands differ in the importance of the role of ice. The far north Canadian islands are encased in permafrost that unfreezes to a depth of five feet each summer. The frozen ground limits the amount of sediment that can be moved about by waves in a big storm. During the winter the islands are locked in ice, unaffected by any and all oceanic storms. During the spring ice breakup or the fall freezeup, blocks of ice forced ashore by winds scour the island and bring in sand and gravel.

According to Taylor, the islands are not building seaward like those in Iceland. They don't seem to be migrating landward, either. At least for this instant in geologic time the Cape Burney sandur islands remain in place.

THE SANDUR TRIBE

The sandur barrier islands of Iceland have perhaps the largest sand supply of any barrier island chain in the world. Even without the occasional jokulhlaup, the Skeidararjokull Glacier produces immense amounts of material, crushed and plucked by the ice. As this vast supply of sand arrives at the shoreline, the shoreline is pulled seaward by the advancing shoreface at around fifteen feet per year. Many barrier islands around the world have at some time in their past widened or been pulled seaward as the shoreface advanced seaward. For these barriers, the seaward advance of the island shoreline is widening, not migration. Only the sandur islands truly migrate in a seaward direction.

The sandur islands demonstrate, as no other island type does, the critical importance of the offshore source of sand. These islands have no other source of sand, no updrift eroding mainland shorelines, no eroding river deltas. Only the shoreface provides the island building materials. The Icelandic islands give positive proof for what we have inferred from beaches everywhere: the surf zone is essentially a long sand tap. It spews sand up to the beaches, sand that is later overwashed or blown into islands, building their elevation. Of course, the shoreface sand originally came from the rivers that flow through the inlets.

If sea level change plays any role in the evolution of these islands, it cannot be detected. Its importance is far overshadowed by the huge waves that straighten the shoreline and by the enormous sand supply. But global warming is playing a role. The accelerated retreat of Icelandic glaciers since World War II has led to larger sand supplies, which results in widened and more stable islands and inlets.

The sandur islands around the world will presumably never see permanent habitations of humankind. Living in a cold climate snuggled at the foot of a glacier is too perilous even by the standards of those who jam the Temperate Zone islands in "Hurricane Alley." The *sandur* barriers will remain hugely dynamic islands, remote and rarely seen but steady in their icy domain.

CHAPTER 10

Arctic Islands

The Cold and Dark Ones

No barrier islands in the world are riskier places to live than those along the shores of the Arctic Ocean (fig. 10.1). Remoteness, severe weather, cold water, cold air, shore ice, and polar bears are among the hazards faced by the Inupiat Eskimos in this completely unforgiving environment. Archaeological excavations in Barrow, Alaska, at the extreme western end of Alaska's North Slope barrier island chain strikingly brought this point home.

After storms, residents of Barrow often stroll along the beach at the base of the frozen bluff in front of their village. They look for artifacts, traces of their ancestors that erode out from the bluff face. The frozen family, as it came to be called, was discovered when a boot, with a foot in it, was discovered protruding from the bluff.

Archaeologists learned that about eight hundred years ago, an ice ride-up occurred in ancestral Barrow. Ice ride-ups take place when wind pushes a large sheet of sea ice, sometimes hundreds of yards wide, onto the land, where it may move inland at the walking pace of a person, for as much as half a mile. Unfortunately this ancient ice ride-up encountered and crushed a house made of skins, driftwood, and sod. Wood in this treeless world is driftwood from rivers as far away as Siberia. The occupants were killed instantly, in the positions they occupied at the moment when their home was flattened. The people were also instantly frozen and preserved, complete with the clothing they wore, the tools they used, and the food they were about to eat. The ice ride-up tragedy provided a unique and most revealing snapshot of Inupiat life, including the fact that the family was suffering from malnutrition.

FIGURE 10.1

A spring scene on a barrier island near Point Lay, Alaska. Three-to-fifteen-foot-high ice-push mounds that have not yet melted line the ocean side of the island. When the ice chunks melt, piles of new sediment will be left behind. *Photo by Andy Short.*

Ice ride-ups and many other hazards continue to be a part of Arctic coastal life. Kotzebue, Alaska, a small town of three thousand people just north of the Arctic Circle along the Chukchi Sea, is threatened by ice ride-up every few springs. The moving ice crosses the narrow beach and slides across the town's main thoroughfare, threatening businesses and residences. The town's bulldozers respond by crumpling and pushing the ice sheets back. Everyone knows, however, that the day will come when the ice sheet will be too large and will move too fast to be held back with mere earth-moving equipment.

As on barrier islands everywhere, processes like ice ride-up that are necessary for the evolution of Arctic barrier islands, are the very events that either brush people off the islands or lead to massive engineering expenditures to halt the processes.

ARCTIC SCHOLARSHIP

In my younger days I was a specialist in the study of abyssal plains, the huge flat areas far from land on the deep ocean floor. It was almost a micro-specialty, one in which all of the world's specialists could fit into a single hotel room to discuss their latest discoveries. Arctic coastal geology is like that. The number of practitioners, all hardy, outdoors-oriented people, is very small.

Somewhere back in the 1960s, the U.S. Geological Survey recognized the need for a greater understanding of Arctic Ocean processes. The smell of oil was just over the horizon. Given the potentially lucrative oil deposits that existed on the North Slope, extensive exploration activities by the oil companies were a certainty. An understanding of how the shorelines and the continental shelf worked would be a necessary precursor to reduce the inevitable environmental impacts of drilling for oil in a pristine wilderness.

So it was that American geologist Erk Reimnitz began a professional lifetime of Arctic studies, often partnered with U.S. geologist Peter Barnes and Arctic experts Bob Taylor and Steve Solomon, his counterparts with the Canadian Geological Survey. Across the Chukchi Sea in Siberia were a few Russian coastal geologists carrying out similar studies. The most notable of these was Usevolod Zenkovitch, who wrote one of the earliest books on coastal geology, which included a number of observations on the Siberian barrier islands.

Pioneering fieldwork in Arctic waters was fraught with peril and excitement. An early misadventure instructed the geologists about the use of their forty-two-foot boat, the *Karluk*, in ice-filled waters. They had anchored for the night in an inlet with a significant tidal flow, and that turned out to be a big mistake. The four-person crew was just sitting down to eat when suddenly the boat lurched heavily. Reimnitz rushed to the deck to discover that a huge ice floe, towering over the boat, had run up against the anchor chain and was pulling the boat under. The boat's propeller was already high and dry and the bow almost submerged. The crew managed, just barely, to push the boat away from the ice, averting frozen disaster and probably death. Thereafter, the group's number one rule was to choose anchorage where ice was unlikely to run the boat down. Although they never again had a serious problem, Reimnitz vividly remembers the *rat-a-tat* sound of small ice floes running up the anchor chain almost every night as they rested, but probably always half awake.

Polar bears were a constant worry as well, and the field parties were always armed with rifles, just in case. Once, while anchored near Norwalk Island, Reimnitz glanced out a porthole to find two polar bears standing at the nearby shoreline, noses in the air, sniffing the odor of a cooking meal that wafted ashore from the geologists' boat. Bob Taylor, the Canadian geologist, was much less lucky. Two polar bears, in two separate incidents, had to be shot to save his life. In one day of terror, he was dragged out of the campsite, his head in the jaws of a young bear, before a companion shot it.

For years Alaska Airlines pilots would take planes into a slight dip to announce the crossing of the Arctic Circle. Thousands of tourists and workers, heading for Kotzebue, Barrow, Prudhoe Bay, and dozens of smaller villages, were entertained by the "Arctic Circle bump," until the humorless Federal Aviation Administration stopped the practice.

The Arctic Circle (60 degrees, 30 minutes latitude) is the line on the globe above which the sun never sets for at least one day (June 21) every year. Situated 1,630 miles from the top of the globe (the North Pole), it is also the line above which the sun never rises for at least one day (December 21).

Smallest of the world's oceans, the Arctic is also the coldest, and most of it is permanently covered by ice. This ocean touches three continents—North America, Asia, and Europe—and is subdivided into a number of arbitrarily defined seas (figs. 10.2 and 10.3). So much ice, so much water, so much cold, the Arctic Ocean generates weather patterns that have a substantial impact on the rest of the globe.

The atmospheric circulation off the North Slope of Alaska, known as the Beaufort Gyre, produces a clockwise flow of wind and water in the Beaufort Sea that forces ice to drift west, either onto or along the shore. This keeps the permanent ice sheet close to shore, which reduces the fetch. Open stretches of ice-free water average less than twenty miles in width at

FIGURE 10.2 *(left)*
Map of the Arctic Ocean showing the seven constituent seas. The darkened patch covering most of the ocean is the approximate location of the permanent ice sheet.

FIGURE 10.3
Map showing the barrier island chains in the vicinity of Prudhoe Bay on the North Slope of Alaska. Cross Island is located a greater distance from the mainland (and closer to the permanent ice sheet) than any other North Slope barrier island.

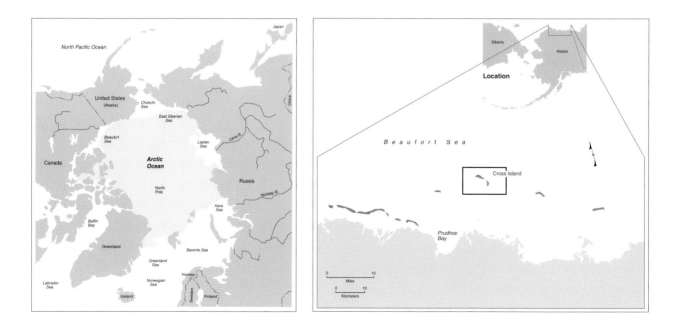

the end of the summer. In contrast, the Alaska Chukchi Sea shoreline has a summer fetch in excess of one hundred miles, a situation that creates much higher waves.

Tides in the Beaufort Sea are extremely small, less than one foot, so on most days wind, not tide, determines the level of the sea. Summer waves arrive from the east or northeast, producing the strong east-to-west long-shore currents that move sand on barrier island beaches and extend sand spits to the west. Because cold water is denser than warm water, waves breaking on an Arctic beach can suspend and carry more sand than waves on a Florida beach. Storm waves are usually less than three feet high.

On Siberia's Laptev Sea, the dominant wind direction is offshore. As a result, ice is pushed away and six-hundred-mile-wide stretches of open coastal water form during the brief Arctic summer. Parts of the Laptev even remain ice-free all winter. With such a large fetch, very large storm waves can form and break on barrier island beaches when the prevailing winds reverse and blow toward the land.

Arctic Ice

Along Alaska's North Slope, spring is a time of violence and change. During the spring ice breakup, the power of the Arctic is on full, awesome display—beautiful sights and strange sounds abound, as the Arctic seems to come alive after its long winter sleep. Cracks begin to form in the sea ice, and seals pop up, keeping a wary eye out for the canny Eskimo hunters. Both the hunters and the seals keep a sharp eye out for hungry, ill-tempered polar bears, especially mother bears with their newly born cubs.

By the time the sea ice begins to crack, the spring breakup of the rivers is already well under way. The rivers tumble to the sea in full flood. They can't vanish into the ocean as would happen in warmer climate regions, so they pour out onto the ice instead and rush seaward as much as twenty miles from the shoreline. The extended river finally spills into the sea when it encounters a crack (strudel) in the ice (fig. 10.4). Roaring whirlpools that are audible for miles in the Arctic stillness mark the point where the river drops into the sea.

The catastrophically descending river water explodes on the seafloor and digs circular depressions, called *strudel scours*, as deep as twenty feet and up to thirty feet across. Simultaneously, tons of sediment are lifted into the water column to be transported away by ocean currents.

If the rushing river water doesn't pour down into a crack in the ice, it may slash across a barrier island and excavate a ditch. Later, after the ice

FIGURE 10.4

A strudel in the fast ice off the North Slope of Alaska. Water from a nearby stream in spring flood flows over the still extant fast ice and spills through the strudel or crack in the ice. The cascading water smashes to the seafloor, forming scour depressions called strudel scours. In this photo it is possible to see the flow patterns of the stream across the ice, all leading to the strudel. A circular thermokarst lake can be seen on the mainland in the upper part of the photo. *Photo by Erk Reimnitz.*

melts, the ditch grows into an inlet. It is a mode of inlet formation completely unique to the far north.

Every year the Arctic barrier islands go through *breakup*, *open water*, and *freezeup*, each characterized by a different set of physical processes. During the two- to three-month phase of open water, the barrier islands and beaches behave much like those in temperate zones, where the principal forces are wind and waves. The one key difference is that the sand and gravel of the islands remain frozen (permafrost) just below the surface, even through the summer months.

Ice comes in many forms. The first, which appears at summer's end, is *frazil ice*. Frazil consists of small crystals that coagulate and form slush in Arctic surf zones when turbulent water is at or below the freezing point. These ice crystals form on the seafloor and then float to the surface, sometimes incorporating sediment grains. *Fast ice*, the immobile ice sheet that forms at the end of the brief Arctic summer, encases the islands and rarely moves about until the spring breakup. In the Beaufort Sea off the North Slope, fast ice grows to a thickness of six feet by the end of the winter.

Seaward of the fast ice, which ordinarily is not more than twenty miles from the Alaska North Slope shoreline, is a huge zone of multiyear *pack ice* that covers much of the Arctic Ocean. Pack ice is a jammed-together mixture of ice of a variety of sizes, ages, and origins. Pack ice covers enormous areas of ocean, with few stretches of open water. In the Beaufort Sea, pack ice, which is constantly in motion in both winter and summer, can be as thick as 250 feet.

A zone of particularly active ice movement exists at the point of contact between pack ice and fast ice. There, sheets of ice buckle into jumbled, irregular ridges that stretch up to several miles in length. Usually the ice ridges, called *stamukah*, don't survive the entire summer—they melt away. When they persevere, however, they act as a giant offshore breakwater that calms the open waters in front of the barrier islands and slows down the summer island evolutionary processes. Cross Island (batik 10.1) is located close to the zone of *stamukah* formation and is often protected by them.

Constant tumultuous activity of Arctic ice makes the shallow ocean bottom in this northernmost region the most roughed-up of any in the world. The keels at the bottom of *ice floes*, single pieces of floating sea ice pushed around by winds, gouge the seafloor. The small channels or *ice gouges* may extend along the seafloor for hundreds of yards. When an ice floe is grounded, it moves up and down and rocks back and forth with the action of waves. It is a motion that produces sediment features that bear a resemblance to miniature volcanic cones called *ice wallows*. They can be as large as a football field.

The bumps and troughs and cones on the Arctic seafloor may last for years. Or they may be obliterated in the same year they formed, by summer or early fall storms that smooth out large areas of the continental shelf. After the storms pass, the processes of gouge, scour, and wallow start all over again the next spring.

Ice rafting augments the sediment transport and mixing effects of all the ice processes. Each ice floe that scrapes, bumps, or gouges the seafloor picks up sediment and moves it over distances ranging from a few feet to miles. A significant portion of the sand fraction of Alaskan Arctic barrier islands was ice-rafted from Canadian waters.

Ice moves up onto the islands in both the fall and the spring. The two major processes by which this occurs are called *ice ride-up* and *ice pile-up*. Ice ride-up, which crushed the house of the frozen family in Barrow, happens when an ice sheet, pushed by winds and pack ice movements, slides smoothly ashore, arching over the contours of the beach and the barrier island. Although spectacular, ice ride-up contributes relatively little sediment to the island.

BATIK 10.1
Cross Island of the Arctic, Alaska, 1998, 30" x 49"
Cross Island, on the North Slope of Alaska, is rapidly migrating toward the mainland and is getting
thinner. It is in its death throes. The white objects in the batik are ice floes that during the Arctic
two-month period of open water, brush past the shallow water in front of the island and bring in or
carry away sand.

Artist's note: Scientist Erk Reimnitz described Cross Island with poetic language as we talked on the
telephone. His writing about the Arctic islands is amazing. Pilkey and I watched hours of aerial
videos and found this image in Erk's aerial photographs. As a silk batik, it feels like a new moon in
a nest of stars, but in reality the pointed ends of Cross Island are shaped by ice and waves.

Ice pile-up is different. Ice floes or small ice sheets break up and come ashore, piling up on the island into mounds of sediment-laden ice (fig. 10.5), usually within a few tens of feet of the beach. Ice pile-up mounds as high as fifty feet have been reported, but ten- and twenty-foot-high mounds are more common. By the time the summer melting is over, piles of sand and gravel remain, often visible from well out at sea. During the next ice pile-up season, remaining mounds from the previous season may block the ice sheet and cause additional pile-up in the same location as before.

Today, most Arctic barrier islands of the Beaufort Sea migrate too rapidly to be dominated by ice pile-up. Erosion rates are so rapid that ice pile-up sediment mounds are obliterated by a retreating shoreline before they can accumulate and multiply from season to season. There is some evidence that a hundred years ago, ice pile-up mounds were more common than they are today, a possible indication of recent increases in island migration rates.

THE NORTH SLOPE ARCTIC BARRIER ISLANDS

All told, 475 barrier islands line the Siberian, Alaskan, and Canadian shores of the Arctic Ocean (appendix, table 3). Polynia Island, a barrier off the much larger and rocky Brock Island, Canada, holds the distinction of being the northernmost barrier island in the world. It is six hundred miles closer to the North Pole than the islands of Alaska's North Slope.

FIGURE 10.5
Close-up of sediment-rich ice-push mounds on a beach near Barrow, Alaska. Perhaps 90 percent of the volume of the mounds is ice. These are similar to the mounds shown in figure 10.1 except that melting is further along. *Photo by Jess Walker.*

The surface of Alaska's North Slope coastal plain is covered by *tundra*, a soil and vegetation assemblage characteristic of dry, cold climates. Mosses, lichen, grasses, and sedges make up the treeless group of tundra plants. Tundra is usually a dark, mucky, gravelly soil that reflects a long history of development in an extreme cold climate. Underlying the vegetation, within a few inches or feet of the surface, is permafrost, or permanently frozen ground.

During the brief summer, shallow-melt lakes form on the tundra near the coast. These *thermokarst*, or thaw, lakes (fig. 10.6) persist because permafrost, which lies below them, is impermeable, so the water can't soak in. Summer winds whip across the thaw lakes and erode their melting tundra shorelines, a process that eventually causes the lakes to expand and join one another. These lakes are central to the development of the North Slope barrier islands.

When only a narrow strip of tundra separates the sea from the elongated lake, storms break through and form a ragged shoreline, from which, under the right conditions, spits can grow and eventually form barrier islands (fig. 10.7).

North Slope islands range in length from one to twelve miles and are usually between one hundred and two hundred yards wide and less than three yards high (fig. 10.8). Throughout the Arctic, island length averages

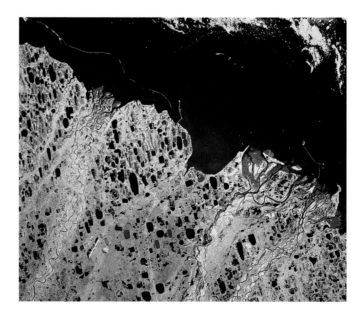

FIGURE 10.6
A late August 1985 image of thermokarst topography on Alaska's North Slope at Prudhoe Bay. Cross Island is in the upper right corner of the photo, very close to the sea ice boundary with open water. The consistent orientation of the long axes of the thermokarst lakes is a reflection of the dominant summer wind direction. The lakes, held in place by impermeable permafrost beneath them, form as the ice melts each summer. After formation, the ponds grow larger each year, as wind waves erode the edges. Breakup of the lakes at the shoreline often results in barrier island formation (fig. 10.7). *Photo furnished by Bruce Molnia and the U.S. Geological Survey, EROS Data Center, Landsat MSS image.*

ARCTIC ISLANDS

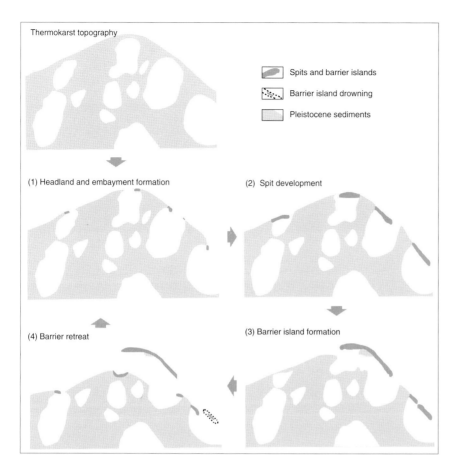

Thermokarst topography

Spits and barrier islands
Barrier island drowning
Pleistocene sediments

(1) Headland and embayment formation

(2) Spit development

(4) Barrier retreat

(3) Barrier island formation

FIGURE 10.7 *(above)*
The sequence of formation of Arctic barrier islands as thermokarst topography erodes. The erosion process concentrates the sand and gravel from the tundra to eventually form islands. It is in some ways similar to the processes that form barrier islands by erosion of Mississippi River Delta lobes. *Modified from Ruz, Hequette, and Hill.*

FIGURE 10.8
A view of the open beach on Thetis Island on Alaska's North Slope. This island is low, has no vegetation, and is dominated by gravel-size sediment, a common type of North Slope Island. *Photo by Jess Walker.*

less than two miles, much shorter than typical islands in the Temperate Zone. Only 12 percent are longer than six miles. Probably the fact that Arctic longshore currents have only a couple of months each year to transport sand and lengthen the islands keeps individual islands shorter than their southern counterparts.

Three stages of life characterize the Alaska North Slope barrier islands. They first form as tundra strips (youth), lengthen through spit formation (maturity) and then begin to disintegrate from starvation through lack of sediment (old age). Most North Slope islands are at the old-age stage (figs. 10.9 and 10.10). Inlets between sand-starved islands tend to be much wider than the islands are long. Most of the islands are migrating and, at the same time, diminishing in size.

Commonly, Arctic barrier islands have strongly curved spits at either end. In the Kara Sea off Siberia (Severnya Island) and in Canada's far north (Prince Patrick Island) the curvatures are so extreme that the islands are circular, almost like coral atolls (fig. 10.11). Crescent-shaped islands may reflect a low sand supply that prevents the normal extension of spits on the end of islands. Perhaps curved islands result also because the maximum fetch for wave formation is often from the sides of the barrier islands rather than from the front facing the ice-covered expanse of the Arctic Ocean.

Perhaps the most famous example of a curved Arctic island is Cross Island (batik 10.1). Called Napaksralik by the Inupiat, it lies at the western terminus of a fifty-mile-long chain of old-age barrier islands. It is the seawardmost island along the North Slope. Never the site of permanent habitation, the island has long been a base for subsistence fishing, hunting, and whaling. Small storm shelters and buildings to store whale meat were built

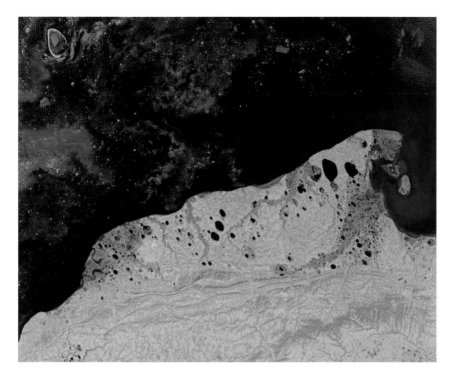

FIGURE 10.11
An "atoll" barrier island in the Laptev Sea off Siberia. Pescheny Island, in the upper part of the photo, is actually two islands, separated by inlets. Thermokarst lakes are visible on the Siberian mainland. Probably such circular islands form where fetch (distance of open water available for wave generation) is similar in all directions during the brief ice-free period in the summer. *Landsat 7 image from the U.S. Geological Survey.*

periodically, only to be knocked down by ice override. Now local Eskimos complain that the mining of gravel, to be used for the oil production facilities in nearby Prudhoe Bay, is making hunting of all kinds difficult.

Cross Island's location made it ideal for whaling, since the spring and fall whale migration from the Bering Sea to Canada's Mackenzie River and back occurs along the edge of the ice pack where the island is located. Today, each of the North Slope villages has an annual whale quota, usually less than five. Some of the whaling is still carried out in traditional skin boats, and if a hunt is successful, the whale is divided up among villagers according to a very complex formula, giving priority to village elders.

The region's extreme shoreline erosion rates hasten the death throes of the Arctic islands (fig. 10.12). The North Slope islands, which erode at rates of six to twenty feet per year, may, as a group, be the most rapidly deteriorating islands in the world (figs. 10.13 and 10.14). During a single storm of unusual intensity, the entire seaward front of Cross Island moved back sixty feet, and the island never recovered. Studies of aerial photos from the North Slope show that a number of barrier islands have completely moved off the footprint they occupied in 1950. All this happens despite the low waves, small fetch, and very short season of open water.

FIGURE 10.12

A barrier island, near Icy Cape, Alaska, completely flooded by a storm. The sea is to the right, and the lagoon is to the left. The small boat is anchored on the submerged crest of the island. The low and closely spaced storm wave crests are typical of Arctic storms where fetch is very small. Because of the closeness of the permanent sea ice to the ocean side of the North Slope barrier islands, the fetch may be greater in the lagoon than in the open ocean. Andy Short, who was wearing a dry suit, took this photo. When asked why he was standing thigh-deep on a flooding Arctic island during a storm with gusts of eighty miles per hour, with a small single-engine boat more than a mile from shore, he said he was protected by the immortality of a graduate student in his early twenties. *Photo by Andy Short.*

Then why the rapid erosion rates? Erk Reimnitz and his coworkers aren't certain of the answer. Something changed very recently, perhaps within the twentieth century. If current rates of island degradation had existed for several centuries, the islands would not endure today. Sea level rise may be involved, but the very limited data suggest that the level of the sea may even be static here at the moment.

One factor must be that these barrier islands don't recover from storms, as Temperate Zone islands do. The seaward loss of beach sand by storm waves is a one-way process in the Arctic. The fair-weather waves, between storms, are not energetic enough to bring all of the storm-removed sediment back to the beach.

Another possible cause of the recent acceleration in island migration rates is suggested anecdotally by the logs of whaling ships. Apparently there is lot more floating summer ice now than whalers saw in the nineteenth century. This change implies that a warming climate is causing more ice to break off in the summer. The transport of sand by ice rafting could vastly accelerate as the larger numbers of drifting ice floes brush against the seafloor off the islands. More open water would also mean more fetch and larger waves. In other words, the Arctic islands may be adjusting to a new set of physical conditions, perhaps brought on by the warming of the Arctic Ocean.

Finally, warmer summers may have increased the annual summertime "melting" of the permafrost underlying the island, the beach, and the ad-

jacent seabed. Frozen beach sediment may be well over 50 percent water. When the ice melts in the beach and in the shoreface, sediment volumes decrease enormously, and that results in a virtual collapse of the seafloor, which in turn causes an immediate leap in erosion rates. All of this, however, is still a poorly understood process.

SARICHEF ISLAND ALONG THE CHUKCHI SEA

Named after an early Russian explorer, the seven-island Shishmaref barrier island chain (fig. 10.15) lies astride the Arctic Circle, adjacent to the Bering Strait. These starkly beautiful, grass-covered islands have no trees or bushes and rim the end of Alaska's Seward Peninsula (batik 10.2). Most lie within the 2.7-million-acre Bering Land Bridge National Preserve. The dynamic nature of these islands falls somewhere between that of the North Slope and the Siberian islands—with their strong dependence on ice—and that of the temperate islands—whose evolution depends on wind and waves.

About six hundred Inupiat Eskimos live in the village of Shishmaref, situated on the four-mile-long Sarichef Island (fig. 10.16), where the principal occupations are whalebone carving, hunting, and fishing. The villagers' diet is sustained mostly by meat: two kinds of seals, musk ox, moose, reindeer (two herds are maintained on the mainland), and fish. They do not participate in the annual Inupiat whale hunt because the whales do not enter the shallow waters off the Shishmaref island chain. The whalebones carved for the tourist trade by the villagers are from carcasses that drift onto the beach.

FIGURE 10.13 *(left)*
The Arctic summer calm before the storm in Tuk, Canada. *Photo by Matt Stutz.*

FIGURE 10.14
The same scene as in figure 10.13. A very large storm by Arctic standards, the waves are three to four feet in height, and the storm surge is about six feet above normal sea level. The storm has removed the tent, and the grounded derelict vessel has been reoriented. The shoreline moved back approximately sixty feet during this storm. As a rule, Arctic shorelines do not recover from storms because the fair-weather waves are too small to bring sediment back to the beach and island. *Photo by Matt Stutz.*

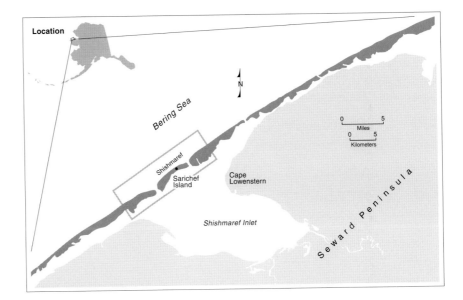

FIGURE 10.15 *(above)*
Map of a portion of the Shishmaref Island chain. The village of Shishmaref is on Sarichef Island.

FIGURE 10.16
Aerial view of Sarichef Island along the Chukchi Sea. Shishmaref village is the dark ground on the narrowest part of the island. *Photo by Owen Mason.*

Four-wheel-drive, all-terrain vehicles bounce along the hard-packed sandy roads in the summertime. When winter and the long darkness come, the four-wheelers are replaced by noisy snowmobiles that shatter the still of the Arctic night. Snowfall is light in this desert climate with only eight inches annual precipitation. The village's centerpiece is a well-maintained cemetery lined with freshly painted white wooden crosses. The religion practiced in this and other Inupiat villages seems to depend on which missionary got there first, a century ago. On Shishmaref, the Presbyterians were the first to arrive.

ARCTIC ISLANDS

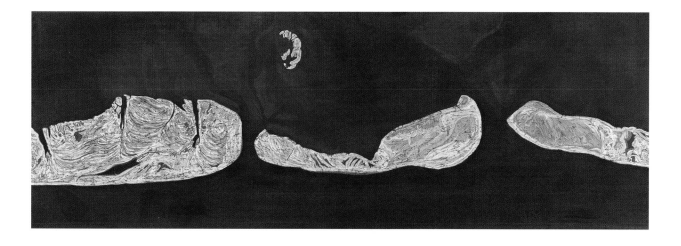

BATIK 10.2
Shishmaref's Shores, Alaska, 1996, 34″ x 98″
Permafrost plays an important role in the evolution of Alaska's Shishmaref Islands facing the Chukchi Sea. These islands have a four- to five-month annual open-water period, but during the winter stormy season they are completely enclosed in ice.

Artist's note: The batik is taken from a high-elevation black-and-white photograph. My art flips the picture as if a plane were flying toward the mainland. The barrier islands become totemic canoes sailing through dark waters. An embryo of hope hovers above the islands for the Native Americans who inhabit this icy shore.

The villagers lead a very isolated existence that falls somewhere between their old nomadic life and the modern American way. One sign of the isolation is the paper money worn white from years of use without replacement. It is also the reason that posters on the walls of "city hall" describe how to behave in Anchorage: don't look passersby in the eye and don't get drunk. By law, no alcohol is allowed, but it becomes apparent that the rule is sometimes circumvented.

The island to the north of Sarichef Island has a spectacular display of storm rings (batik 10.2), the arc-shaped half circles of sand formed by overwash of successive storms. Similar storm rings are formed on other islands in Alaska (fig. 10.17), as well as on the Outer Banks of North Carolina.

Except for the lack of bushes and trees, the Shishmaref islands are similar in overall appearance to some of the narrow islands along the East and Gulf Coasts of the Lower Forty-eight. The dunes are held in place by dune grass, and extensive salt marshes border the lagoon shoreline. During the summer, beach sand moves Arcticward to the northeast under the influence of the wave-driven longshore currents. The Chukchi Sea is free of ice for four to five months of the year, in contrast to the North Slope, where two to three months is the open-water norm. Ice ride-up, ice push, and all the myriad processes that churn, mix, and plow the seafloor and beaches further north are less important here. In addition, the winter fast ice here is thinner and shorter-lived, and the permanent ice pack is many miles to the north. As a result of the duration and the fetch of Chukchi Sea open water, more time and wave energy are provided for the islands to evolve and lengthen in tune with the winds and the waves. On average, Chukchi Sea islands are seven miles long compared to an average two miles in length along the rest of the Arctic shorelines.

FIGURE 10.17

Storm rings formed by cross-island overwash on a barrier island along Kotzebue Bay, Alaska. These interesting features are found on the Shishmaref Islands, Alaska, and on Core Banks, North Carolina, as well as other places. *Photo by Miles Hayes.*

Owen Mason, an Anchorage geologist and student of the Chukchi Sea islands, uses the accounts of early explorers to document changes in barrier islands. Alexander Kashevarov, a Russian who explored the Alaskan Chukchi coast in 1838 using shallow-draft skin boats, observed and carefully recorded a number of barrier island inlets that no longer exist. Mason's own observations indicate a similar recent decrease in inlet numbers. He notes that on the Shishmaref island chain there are a total of twelve relict flood tidal deltas that extend into the lagoon but are no longer connected to an inlet. The reason for the decreased number of inlets (and increased island length) in the last couple of centuries is not certain. The Outer Banks of North Carolina experienced a similar decrease in the number of inlets in the same time frame. In both cases, a decrease in the number of inlet-creating storm events may be responsible.

The entire Shishmaref barrier island chain is within Alaska's permafrost zone. "Hardened" ground minimizes the immediate impact of storms, in that storm erosion comes to a near halt as soon as permafrost is exposed by the crashing waves. Like most of the world's barrier islands, the shorelines of the Shishmaref chain are slowly retreating, but at nowhere near the speed of shoreline erosion on the North Slope islands.

Still, erosion has long been a problem, say the Shishmaref village elders (fig. 10.18). And it has increased dramatically in the last decade. Ironically, the old village of Shishmaref, in the sixteenth to nineteenth centuries, was

FIGURE 10.18
The beach in front of the village of Shishmaref. The beach in front of the seawalls, some of which have collapsed, has narrowed to the point that the waves break on the walls each high tide. *Photo by Owen Mason.*

set on dunes, well back from the shoreline, in a location less threatened by erosion and storms than the present village site. The village was moved to its current, more dangerous site to facilitate offloading of supplies, including government-issue houses, from barges grounded on the ocean beach.

Usually the sea and the lagoon freeze over in October, the same time that the big storms begin. For a period of about one month the islanders are isolated because the lagoon is too frozen for boats and not frozen enough for snowmobiles and dogsleds. In 1997, during El Niño, the winter freezeup did not occur until November. As a result, the shoreline remained vulnerable well into the storm season, and serious erosion threatened a number of buildings and small oil tanks. During the storms, inhabitants emptied their yards of old dogsleds, snowmobiles, and other abandoned household goods and threw them into the surf zone in a hopeless attempt to hold back the sea. The permanently frozen island sand probably minimized erosion and prevented a disaster.

Using junk to stabilize shorelines is an old Chukchi Sea tradition. On Teller Spit near Kotzebue, World War II surplus trucks and other military equipment that somehow found its way to the remote Arctic, line the beach, along with concrete-filled fifty-gallon oil drums. At Point Hope, a boulder seawall on the beach protects caches, excavated into the permafrost and used by the Inupiats for food storage. The eventual and inevitable negative impact of the wall on the sand supply of adjacent beaches seems a stiff price to pay to protect simple and easily replaceable excavations for caches.

Long-term plans include the possibility of moving the entire community, with its airstrip, to the mainland, a costly $50 million exodus across frozen land. In the old days of a mobile, flexible existence, moving a village was easy. Even as recently as the 1970s, Point Hope was moved back, two miles away from an eroding shoreline bluff. Now water supply, sewage disposal, and permafrost construction requirements mean that the relocation of even a small village would turn into a major construction project.

THE COLD ONES

The Arctic barrier islands line 24 percent, or 1,635 miles, of the long shoreline of the Arctic (appendix, table 1). As measured in terms of length, they are divided as follows: 55 percent are Russian, 23 percent are Alaskan, and 12 percent are Canadian. Measured by length, the 569 barrier islands make up 12 percent of the world's barrier islands (appendix, tables 2 and 3). The global percentage as measured by the number of islands, however, is a much larger 22 percent. The disparity between the importance of the Arctic islands in terms of number and length has to do with the fact that the Arctic barriers are short, around two miles long on average. This is because only a brief time is available during the Arctic summer for waves to lengthen islands. The Chukchi Sea islands, which have at least three times as much fetch and twice as much ice-free time, average six miles in length.

Other Arctic–Temperate Zone differences are the prominence of ice-related processes, the importance of gravel in some barriers, and the lack of a role for vegetation in island evolution processes (except for the Chukchi Sea islands). The biggest storms to strike the Arctic coasts have no impact because they occur in the winter, when the islands are locked in ice.

Although the Arctic islands are governed by the cold climate, they share some similarities with barrier islands the world over. Even in the far north humans have mined the beaches and turned to shore hardening when erosion threatens. These islands exist in a natural equilibrium, as do other barriers. When the natural balance is tilted, the islands respond.

The Arctic islands appear to be a highly sensitive barometer of the temperature rise of Earth's atmosphere. A warming ocean provides the best explanation for the ongoing rapid migration and degradation of the Arctic islands. Unlike the rest of the world, rising sea level may not be a major cause of increased barrier erosion rates here. Increased summer wave fetch, greater numbers of small ice floes, and permafrost melting on beaches and shorefaces may, in varying combinations, be responsible for Arctic island activity. Increased fetch leads to bigger waves, and increased numbers of ice floes lead to more sand removal by ice rafting from the islands.

The North Slope islands and perhaps most other Arctic island chains have too little sand to maintain themselves in a warming ocean. They are destined to disappear in a century or so. Not to worry, however. New barrier island chains can be expected to form as open-sea waves begin breaking directly on the thermokarst topography of the mainland.

CHAPTER 11

False Islands

Things Aren't What They Seem to Be

GEORGIA'S STRANDED ISLANDS

I first suspected that there was a serious problem with my doctoral dissertation when I began to learn how barrier islands worked. I was a newly minted Ph.D. on my first job (1962) at the University of Georgia Marine Institute on Sapelo Island, and I had just (proudly) published my dissertation in an international journal. The study demonstrated the relationship between water temperature and the chemical composition of several species of seashells and noted that it might be possible to determine the water temperature of ancient oceans using shell composition. The shells for my study were collected from cold water to warm water on beaches from Maine to Puerto Rico, and I assumed that they had grown in the waters just offshore from where they were found and that they had recently died.

What I observed on the beaches of Sapelo and other islands in Georgia made me realize that my basic assumption was dreadfully wrong. On the beaches were salt marsh mud layers and abundant oyster shells (fig. 11.1) that had to come from the lagoon side of the islands. How did lagoon shells get on the ocean beach (fig. 11.2), and how old were these shells?

I now know that most seashells on southeastern U.S. barrier island beaches are fossils, deposited under water temperature conditions quite different from those of the present day. In fact, this may be true for most barrier islands around the world. What I didn't know when I collected the shells for analysis was that barrier islands migrated and that shells on the beach often spent many, perhaps thousands, of years encased in mud or sand beneath the island as it migrated over them. *Until my eyes were opened on Sapelo Island, barrier islands to me were piles of sand where seashells could be found.*

FIGURE 11.1 *(above)*

Outcropping marsh mud on the ocean side of Cabretta Island, one of the Holocene islands attached to Sapelo Island. The light-colored circular patch in the foreground consists of oyster shells, some preserved in the same position they grew in in the lagoon before the island migrated over them. On Cabretta Beach, as in most southeastern U.S. beaches, the seashells are mostly fossils, hundreds to thousands of years old. Note the fallen trees on the beach in the background. *Photo by Jim Henry.*

FIGURE 11.2

This photograph, taken a few days after the passage of Hurricane Hugo, shows an extraordinary broad mudflat of former salt-marsh sediment exposed on Sandy Point at Cape Romain, South Carolina. This island is retreating at rates of more than thirty feet per year, a process that brings a large number of lagoon seashells to the open-ocean side of the barrier island. The age of the shells on the beach depends on the rate of island migration—the faster the migration, the younger the shells. *Photo by Miles Hayes.*

During the ice ages, every time the sea level reached a peak and began to subside again, a barrier island chain was left behind, a bump on the coastal plain. At least seven former barrier island chains are visible on the coastal plain of Georgia, each marked by sand ridges (fig. 8.5). The most recent sea level high before the present one, perhaps 120,000 years ago, peaked at about the same elevation as the present sea level. This means that two barrier island chains, one ancient and one modern, are in very close proximity to each other. On coastal plains all over the globe, the close conformity of the two shorelines has shaped the barrier island coasts. Sometimes the two islands combine to form a single large island, made up of both modern and ancient islands, giving the false impression of a single island with a large sand supply.

Certainly the best-known North American examples of composite barrier islands are the islands of Georgia. The eight-barrier-island chain in the state of Georgia and one island each in South Carolina and Florida make up what are locally referred to as the Sea Islands. Beginning in the north and moving south, the main island names are Hilton Head in South Carolina; Tybee, Wassaw, Ossabaw, Saint Catherines, Sapelo, Saint Simons, Jekyll, and Cumberland Islands in Georgia; and Amelia Island in Florida. The islands are backed by wide and spectacular salt marshes locally known as the marshes of Glynn (celebrated in the poetry of Sidney Lanier).

Most of the Georgia barriers consist of two islands (fig. 11.3, batik 11.1). The largest, widest, and highest is a Pleistocene island, seaward of which are Holocene islands, sometimes not more than a few tens of yards wide.

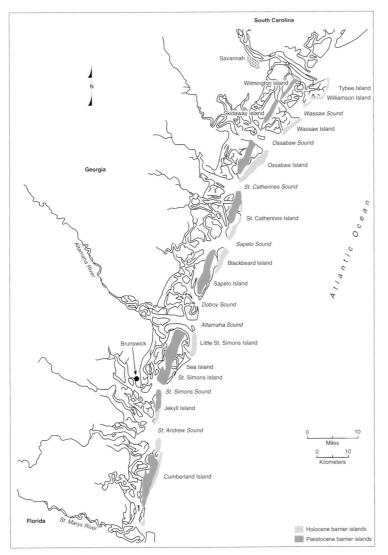

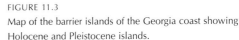

FIGURE 11.3

Map of the barrier islands of the Georgia coast showing Holocene and Pleistocene islands.

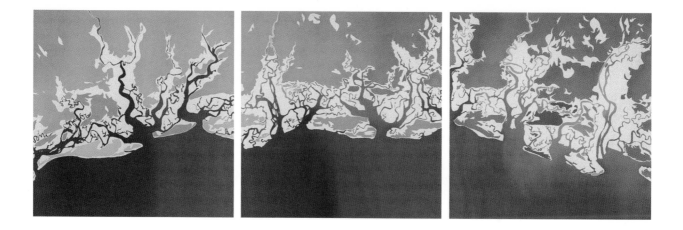

Sea Islands of Georgia, 1994, 48" x 144"
The Sea Islands of Georgia are, for the
most part, combined Pleistocene (Ice Age)
and Holocene (modern) islands. The light
areas on the batik, between the islands
and the mainland, are vast salt marshes
that, except for tidal channels, completely
fill the lagoons.

Artist's note: Sea Islands of Georgia was
commissioned for the Savannah Science
Museum. Nautical charts of the rivers
were used for the motif. If a pelican flew
toward shore at high altitude, this would
be her arriving view. The batik resembles a
group of bonsai trees growing on the top
of a round Earth with branches silhouetted
against a fading sunset.

On portions of Jekyll and Cumberland Islands, the Holocene islands are
entirely gone, and the ocean beach resides on the Pleistocene island. On
Sapelo Island, three small islands—Nannygoat, Cabretta, and Blackbeard
(fig. 11.4 and batik 11.2)—represent the Holocene. Most of the Georgia
Holocene islands have mud layers on the beach, which indicates that they
are actively migrating islands.

Numerous fallen trees and stumps are found on the beaches, especially
on Ossabaw, Wassaw, and Saint Catherines Islands (fig. 11.5 and batik
11.3), a clear indication of rapid shoreline retreat rates. One large tree that
eroded out onto the beach on Cabretta Island remained standing for years,
supported by the remnants of the lateral roots. The still-standing tree was
moved more than a mile down the beach by surf zone currents before it
finally toppled over.

At the north end of the Georgia coast, off the mouth of the Savannah
River, the Holocene islands are widely separated from the Pleistocene islands
(fig. 11.3). The separation (batik 11.1) reflects delta formation at the Savannah River mouth. In other words, the intervening space (now covered by
salt marsh) between the ice age and modern islands is river sediment.

The Pleistocene islands are made up of row upon row of beach ridges that
indicate seaward widening of the islands back when they were active 120,000
years ago. Each beach ridge is assumed to represent a former shoreline position. Why are the Ice Age islands so large compared to the Holocene islands?
The answer remains a mystery, but there are several possibilities. One is that
the sea level 120,000 years ago stayed in one place for a long time, allowing
sand to be pumped ashore by the waves and blown into the dunes for thousands of years. There may also have been differences in rainfall, river flow,

FIGURE 11.4.
Cabretta Island in the foreground and Blackbeard Island at the top of the photo are Holocene islands separated by a salt marsh from the Pleistocene Sapelo Island to the left. The Holocene islands are migrating (note the overwash fans on Cabretta) and eventually will combine with the much larger Sapelo Island. *Photo by Jim Henry.*

BATIK 11.2
McQueen Inlet, Georgia, 1981, 29" x 64"
McQueen Inlet on Saint Catherines Island is a meandering salt-marsh creek, visible at the top of the batik. This small creek drains the area between the modern and the Pleistocene islands. Saint Catherines Sound is in the foreground.

Artist's note: Saint Catherines Island is south of Savannah, Georgia. This oblique angle of McQueen Inlet is one of the first batiks in my "Islands from the Sky" series. In November of 1980 my brother and I were flying in a Cessna 172 at about 1,700 feet when we saw this dramatic sand spit and meandering waterway. Accurate as a landform, the art is simple and spontaneous, like a Japanese painting.

wind patterns, and the amount of sand contributed by rivers at that time. There may have even been differences in the size of tides and waves.

The tide range on the Georgia coast is high: six to nine feet for normal tides and in excess of ten feet for spring tides. This compares with two-foot tides at Cape Canaveral, Florida, and three-foot tides at Cape Hatteras, North Carolina, both at the ends of the deep Georgia embayment formed between Cape Hatteras and Cape Canaveral. The embayment or inward curve of the shoreline between the two capes makes the Georgia coast the westernmost extremity of the East Coast shoreline and puts it out of reach of most hurricanes, which tend to pass it by and hit further north. The continental shelf is eighty miles wide, by far the widest Atlantic shelf in the southeastern United States. Friction between waves and the seafloor across this wide shelf reduces the size of the waves arriving from deep water, providing the Georgia coast with the lowest waves on the eastern coast of North America.

The sand grains are finer on the beaches of the Georgia barrier islands than the sand grains at Capes Hatteras and Canaveral, which very likely reflects the energy or height of the waves. Other things being equal, the higher the average waves, the coarser the sand that resides on the beach. The highest waves, coarsest sand, and narrowest continental shelves are at the capes, and the lowest waves, smallest sand grains, and widest shelf are found off the Georgia barriers.

BATIK 11.3
Blackbeard Island, Georgia, 1981,
33" x 56"
A salt-marsh creek between Sapelo Island and Blackbeard Island. Some of the intricate channels of this marsh creek have been abandoned, leaving behind "oxbows."

Artist's note: I was living in Savannah when we flew our first photographic journey over the Sea Islands. Batiked on raw silk, the serpentine channel river has a timeworn quality. The curious loop in the center and the salt marshes were shot at a low altitude. It is fun to make your living out of thin air with your brother or father by your side flying along at 75 miles per hour with the wind in your face.

As a result of the large tidal range, all kinds of things happen. The beaches are extremely wide at low tide. On Sapelo Island the width can exceed a hundred yards. The great beach width is also abetted by the fine-grain size of the sand. The global rule of thumb is that the finer the sand, the flatter (and wider) the beach between high and low tide lines.

On a larger scale, the tidal range is largely responsible for the close spacing of inlets or the shortness of the islands. Most Georgia islands are less than ten miles long. The high frequency of inlets reflects the large volume of water that comes in and out of the Georgia estuaries because of the high tides and river flow.

Huge tidal deltas that extend far seaward are one more outcome of a large vertical difference between high and low tides. The ebb tidal deltas here are larger than any others on the East Coast and extend seaward as much as eight miles. This compares with seaward extensions of less than a mile for most tidal deltas in North Carolina. The low wave energy also is a factor that helps the tidal deltas extend seaward; there are few high waves to eat away at the rim of the delta. There are no flood tidal deltas because all available sand is stored offshore.

Island People

By the early part of the twentieth century, most of the barrier islands of Georgia were owned by a veritable who's who of wealthy Americans. A twenty-first-century fringe benefit of this era of rich island ownership has been the islands' preservation in undeveloped or lightly developed states. Only Tybee, Saint Simons, and Jekyll Islands have the intense development so common on islands in neighboring Florida or South Carolina. The remaining islands are in private, federal, or state ownership as wildlife refuges, national seashores, state parks, and research and marine sanctuaries.

The Georgia barrier islands were once heavily forested, but during the plantation era large areas were cleared for cultivation. Live oak, a major component of barrier island maritime forests from North Carolina through Florida, proved to be in great demand for ship hull timbers from the mid-1700s until the ironclad era of the mid-1800s. Live oak is a very dense, durable hardwood with a twisted grain capable of deflecting cannonballs. *Old Ironsides*, a War of 1812 warship, was a live oak vessel, its name a reflection of its tough live oak hull, derived from Georgia barrier island trees. Structural components such as curved keel timbers and ship ribs and planks were cut out, taking advantage of the natural curvature of the tree limbs and trunks. Between four hundred and seven hundred large trees were needed per ship, and that made a major dent in the slow-growing, spectacularly beautiful live oak forests of the barrier islands.

Georgia has a far higher percentage of seawall-stabilized developed shoreline than any other coastal or Great Lakes state. This state of affairs came about because of Hurricane Dora in 1964. My family and I were on Sapelo Island when it struck. Dora was not impressive in terms of winds, flooding, or wave height. In fact, my most vivid memory of the storm is the consternation that I caused when I got one of the university Jeeps stuck on the beach hours before the waves came ashore.

The reason I was out in a Jeep was to survey cross-beach profiles before the storm and compare them with profiles taken after the storm. I established "permanent" markers well back from the beach so I knew where to return. After the relatively mild storm, most of the permanent station markers had washed away—another lesson learned.

The amount of erosion by Hurricane Dora was impressive, at least by Georgia standards. Vice President Lyndon Johnson flew over the Georgia coast, surveyed the damage, and declared that he would alleviate the suffering of the people by bringing a seawall to Georgia. And bring it he did. Most of the open-ocean shorelines of Saint Simons and Jekyll were stabilized with a boulder revetment, to be forever known as the LBJ wall (fig. 11.6). Ironically, the rock revetment was put in before the beach had recovered from the storm, and large parts of the Jekyll Island wall were soon covered by dune sand. With each passing year, however, more and more of the wall is exposed, with an ever-narrowing beach in front of it. The LBJ wall combined with private seawall construction on Sea Island (part of the Saint Simons barrier island complex) and Tybee Island sealed the future of Georgia's beaches. Either they will disappear in front of the wall or they will have to be nourished in perpetuity.

Sapelo Island

Ten-mile-long Sapelo Island was my family's home for three years. My wife, who grew up on an island in Alaska (Petersburg), loved it. After three years I was suffocated by it. I couldn't stand the isolation and the too close living with a very small number of people. Even the incredible beauty and the long beaches where one could be assured of finding no one else for a whole day didn't improve my outlook. Mine was not an uncommon response among the scientists who lived on the island.

At the time, thirty-five people associated with the University of Georgia Marine Institute and the administration of the R. J. Reynolds estate were at the island's south end. Two miles away, several hundred descendants of plantation owner Thomas Spalding's slaves eked out a poor and largely subsistence existence in Hog Hammock. The early 1960s were still the days of racial segregation in coastal plain Georgia, and the R. J. Reynolds plantation managers insisted that contact between the races be

FIGURE 11.6
An aerial shot of the LBJ Seawall on Jekyll Island shortly after it was constructed in the mid-1960s. The wall was built after Hurricane Dora (1964), along most of the developed shoreline of Georgia, leading to eventual loss of much beach, two to three decades later. *Photo by* Jim Henry.

on a working basis only. Naturally, students paid little attention to this edict, and integration was rapidly approaching when we left the island to move to Duke University.

The history of Sapelo Island is perhaps typical of the Georgia Sea Islands. Native Americans inhabited the island at least 6,000 years ago. The most spectacular remaining evidence of their presence is a large shell ring, ten to twelve feet high and one hundred yards in diameter. It is 5,800 years old and made up mostly of oysters but includes other edible mollusk shells as well. It is believed to have served some ceremonial function.

The Spanish arrived in 1573, named the island Zapala (which evolved into "Sapelo"), and established a Franciscan mission to save the Indians. Apparently the Creek Indians were not grateful for the attention, as they eventually revolted and killed most of the priests. The Spanish abandoned the island in 1686 as a result of increasing pressure from the English, who were drifting to the south. A hundred years later, the island, now firmly a part of the state of Georgia, was purchased by French nobles who were fleeing the guillotine of the French Revolution. Between 1789 and 1802, the French refugees' tumultuous life, punctuated with pistol duels, led to the breakup of their colony, to be followed by a cotton and sugarcane plantation owned by Scotsman Thomas Spalding. He became a famous agriculturist, experimenting with crop rotation and plant hybridization. Long-fibered Sea Island cotton was in great demand, and the high profits fed the grand lifestyle of plantations served by many slaves. By the 1830s, plantation life began to go downhill when the soil became heavily depleted, cotton prices declined, and the boll weevil started decimating cotton crops. The Civil War finished the plantations on the Georgia islands, for Union soldiers destroyed most of them. After the war many freed slaves remained on the island, but it was a very poor life, characterized by food scarcity and outbreaks of malaria and yellow fever.

One industry that did thrive right after the Civil War was lumbering. Logs from the mainland were floated downriver to nearby Darien, on the mainland, to be milled into lumber and loaded aboard sailing freighters, bound for Europe. The freighters were in ballast when they arrived in Sapelo Sound, and before taking on lumber they dumped their ballast stones, forming small "ballast islands" next to the ship channels. These are still visible around Sapelo Island today, complete with some plants of European origin (fig. 8.7).

From the Gay Nineties through the Roaring Twenties and up to World War II, the Georgia barrier islands were taken over by some of America's wealthiest capitalists. A colony for the very rich was formed in 1886 on nearby Jekyll Island. By one estimate, one-sixth of the world's wealth resided there, eventually including the Morgan, Rockefeller, Goodyear, and Carnegie families. Howard Coffin (Hudson Motor Company), to be followed by R. J. Reynolds of tobacco fame, purchased Sapelo. Two presidents, Calvin Coolidge and Herbert Hoover, visited Sapelo Island during this time; years later, Jimmy Carter visited Hog Hammock for a church service during his presidential tenure. In the late 1950s, the barns, stables, and other outbuildings of the Reynolds mansion were converted to laboratories of the University of Georgia Marine Institute, where I began my career as a marine scientist.

So the Georgia barrier islands are fake barrier islands, or at least the Pleistocene components are. They have every appearance of modern barriers except they are not in equilibrium with today's waves and currents and winds. They were barrier islands at one time, and they appear to be barrier islands now, but their appearance is an accident of sea level.

ACCIDENTAL ISLANDS DOWN UNDER

In Australia there is a different kind of false island chain, one where the islands never were barrier islands but instead are erosion remnants of preexisting land areas. They look like barrier islands, they are made up of the same materials as barrier islands, but they bear no relationship to the waves of today except that they are eroded by them.

Originally, most of the settlements of the Australian Aborigines were concentrated along their coasts. Like their New World counterparts, they were driven inland by the coming of the Europeans and were forced to live in the often less desirable interior of the continent. During most of their forty-thousand-year tenure on the island continent, up until about six thousand years ago, sea level was much lower than it is now. Most of Australian Aboriginal history must be recorded on what is now the submerged continental shelf.

A vast dune field covered the Queensland coastal plain in those times. The sand was blown ashore, across the beach, and then distributed inland by the strong winds from the southeast. These dune fields, among which the Aborigines lived, became the nuclei of the Australian barrier island chain of today.

The remnants of this dune field, exposed in spectacular five-hundred-to one-thousand-foot shoreline cliffs, make up the barrier islands. The dune sands are composed of layer upon layer of different-aged dune fields that formed throughout the Pleistocene. There were probably eight distinct periods of dune formation, the oldest of which occurred around 700,000 years ago. Layers of ancient soil indicate that periods of forest development intervened between surges of dune building. Australian geologists believe that the stratification or layering in the ancient dune sands reveals that the winds came from the southeast, just as they do today. The dunes were never part of a barrier island system.

The soil horizons at the boundaries between dunes of different ages often bottom out with a weakly cemented dark brown sand layer (easily broken up with a rock hammer), held together by organic matter leached by rainwater percolating downward through the sand above. The layers break up and form "coffee rock" when erosion catches up with them (fig. 11.7). Boulder-size chunks of coffee rock are a widespread occurrence on local beaches. Coffee rock is occasionally found on southeastern U.S. Atlantic beaches.

Six thousand years ago, according to Australian geologists, the sea level rose to its present level and has stayed there ever since. (Like our understanding of most of the world's local sea level histories, our understanding of Australia's will certainly be modified by future research.) The islands of Australia's Queensland Province must have formed at that time and have grown ever smaller over the last few thousand years (fig. 11.8).

It was only natural that Aboriginal legends would evolve around the islands. The Aborigines who lived at the coast were as fascinated by the beauty of the islands as present-day visitors are. One of the myths explains the striking red, yellow, purple, and black color (the principal colors of Aborigine art) of the sandy cliffs facing the sea on Australia's barrier islands (fig. 11.9): Yingie, the spirit of the gods, was a rainbow. One day Yingie got into a serious fight over a woman. When Yingie lost, the rainbow spiraled down and crashed into the cliffs. His spirit then permeated the cliffs, coloring them forever.

From north to south the islands in the state of Queensland are Fraser, Cooloola, Moreton, North Stradbroke and South Stradbroke (fig. 11.10). North and South Stradbroke were once a single island, but apparently a new inlet formed after explosives taken from a grounded ship were detonated on the island in 1894. The blast lowered the elevation and devegetated an area where the island was already very thin. By 1898 natural processes finished the job and formed a new inlet, which still exists today. Extensive mangrove forests behind the islands line the shorelines of the lagoon. Australia has a bewildering array of mangrove species, thirty-seven in all.

FIGURE 11.7 *(above left)*
Close-up of heavily iron-stained dune sand on a Cooloola Island bluff. The sand has been lightly cemented by organic carbon and perhaps iron, forming "coffee rocks." These rocks are common objects on the beaches of the Australian Islands. *Photo by Andy Short.*

FIGURE 11.8 *(right)*
Satellite image of North and South Stradbroke, Moreton, and Bribie Islands, near Brisbane, Australia. For the most part, these islands are not products of the present-day oceanographic climate but rather are erosional remnants of an ancient coastal dune system. A flood tidal delta has developed between North and South Stradbroke Islands, but high waves and a steep continental shelf have prevented formation of an ebb delta on the open-ocean side of the island.

FIGURE 11.9 *(lower left)*
Rainbow Beach on Cooloola Island. The color of this ancient dune sand is derived from dissolution and reprecipitation of iron minerals within the sand. *Photo by Andy Short.*

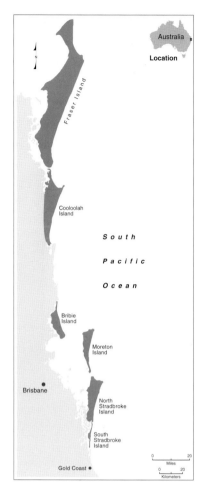

FIGURE 11.10

Map of the Australian islands between Gold Coast and the Great Barrier Reef.

Gold Coast, Australia's Miami Beach, begins just south of South Stradbroke Island. None of the islands is developed to any significant degree, although sand mining has been quite extensive. There were small logging communities on the islands in the late 1800s, harvesting the valuable kauri pines.

Fraser, the largest and best known of the islands, is 90 miles long and said (by Australians) to be the largest sand island in the world (batik 11.4). Maybe it is. Padre Island, Texas, at 135 miles in length, is much longer, but Padre is also much narrower and lower in elevation. Fraser and Cooloola Islands are large enough to have streams on them—a most unusual barrier island feature. Usually rainwater soaks into the sand rather than forming streams. Between the two islands, they also have forty lakes.

The Australians recognize two kinds of lakes on the islands. One is the "standard" barrier island lake or pond, at low elevation, where the ground surface dips below the groundwater table and the depression fills with groundwater. These are referred to as *window lakes*. Shackleford Banks in North Carolina and Bazaruto Island in Mozambique have such lakes. *Barrage lakes* form when a sand dune dams a stream. These unusual lakes are often perched well above the groundwater table, held there in the porous island sand by impermeable peat layers that underlie the lakes (fig. 11.11). Lakes Boomanjin and Boomerang on Fraser Island are at elevations of four hundred feet above sea level. Barrage lakes are also called perched lakes.

Fraser Island was first "discovered" and recorded by Captain Cook in 1770. It is likely that the Portuguese actually sailed by it, and perhaps landed, more than a century earlier. The island was named after James and Eliza Fraser, who were shipwrecked here in 1837, a very early date by Australian settlement standards. The island's Aboriginal inhabitants killed the Frasers and the ship's crew.

The most important factor controlling waves and tides on the Queensland coast is the narrowness and steepness of the continental shelf. The Queensland shelf is fifteen to thirty miles wide, tides are small, and waves are high. Waves come dominantly from the great Southern Ocean and are quite energetic, four to seven feet high being typical. The big storms are tropical cyclones. Shorelines have retreated as much as 250 feet in a single cyclone, although most of the lost sand is recovered within a couple of years.

Tsunamis are the most spectacular of all waves that strike this coast. These occur rarely, but in the tectonically active Pacific Basin, a tsunami is always a possibility. According to studies to the south, in New South Wales, a huge tsunami struck the east side of Australia about 105,000 years

ago. This tsunami eroded rocky slopes a full 50 feet above the present sea level, and since the sea level at that time was below current levels, the runup may have been much larger than 50 feet. The timing seems right for this tsunami to have been a product of the spectacular Lanai mega-landslide that produced a wave runup of 1,200 feet on the island of Lanai in the Hawaii chain. Such a tsunami must have been felt all over the Pacific.

Islands that are considerably wider at one end than the other are a very common occurrence around the world. As we have seen, the drumstick island phenomenon is present on many barrier island chains, including the Dutch and German Frisian Islands, the Portuguese Algarve Islands, the South Carolina islands, the Chukchi Sea barriers, the Niger Delta islands, and even the sandur islands of Iceland.

In a true drumstick, the wide, meaty end is the updrift end of the island, and the narrow part is a spit extending downdrift, in the direction of the dominant transport of beach sand by longshore currents. The rule doesn't

BATIK 11.4
Australia's Fraser Island, 1999, 27″ x 40″
Fraser Island is not a true barrier island, in that it is an erosion remnant from a huge ancient dune field rather than a bar of sand molded by waves, currents, and wind. The batik shows a spit that was produced by longshore currents using sand supplied from the eroding ancient dunes.

Artist's note: Australia's Fraser Island is based on a photograph of this famous island, found on the Internet by Sharlene Pilkey. The subtropical beauty of rain forests beside cliffs of colored sand make this the epitome of an aerial landscape of a restless ribbon of sand.

One of numerous perched ponds on Fraser Island, called barrage lakes by Australians. Most barrier island ponds are at groundwater level, but because of impermeable clay and organic layers in the sand, some Fraser Island ponds, such as this one, are perched well above the groundwater table. *Photo by Alison Smith.*

work on the Queensland, Australia, islands, however; there the fat part of the drumstick is at the "wrong" end. Large quantities of littoral sand (a million cubic yards per year on Fraser Island) flow from south to north along the beaches and from island to island. The widest island expanses are at the north downdrift end.

This strange state of affairs, the fat end in the wrong place, happens because the islands are purely erosional in origin. These were not formed by the deposition of washover fans and sand dunes. Instead they are the eroded remnants of the huge Pleistocene dune field. The northern end of each island is anchored to bedrock headlands, responsible for both the locations of the islands and their "reversed" shape. The headlands provide erosional shelter, or an erosion shield, behind which the sand is protected from erosion. Where there is no erosion shield, erosion is more extensive and the islands are thinner.

A couple of small Holocene barrier islands are found among the Queensland islands, constructed of sand eroded from the old dune field. One is just offshore of North Stradbroke Island, and the other is at the south end of Fraser Island. At the north end of Fraser Island the Holocene sands are draping over the Pleistocene dunes and spilling inland (fig. 11.12) exactly like the Holocene sands on the Mozambique islands. In Mozambique, however, the Holocene sand is climbing over a genuine Pleistocene barrier island dune; that is not the case on Fraser Island.

The sand, which makes up the dunes, ultimately came from river mouths south of the Australian islands. The evidence for this is in heavy minerals contained within the dune sand. Careful study of these minerals, which rarely make up more than 10 percent of the total sand, makes it possible to pinpoint specific source rivers.

Geologists are not convinced by the Aborigine legend of Yingie, the lovestruck god. Instead, they believe that heavy minerals are responsible for the bright coloration of the old dune sand, color that makes a spectacular display on seaward-facing cliffs. The heavy mineral fraction is much more unstable than the quartz and feldspar in the sand, and this instability results in the destruction and dissolution of the minerals during the weathering process. Iron is released to form various kinds of oxides on the surfaces of quartz grains. giving rise to the bright yellows and reds on the cliffs. Manganese oxides perhaps are responsible for the purple colors.

The heavy minerals on all of the Queensland barriers have attracted the attention of mining companies. Three minerals—zircon (for zirconium), ilmenite, and rutile (for titanium)—have been mined from the beach and

FIGURE 11.12
The ancient dune sand that makes up Fraser Island has been remobilized and is forming new dunes. Sand moving ashore from the beach on Fraser Island may have killed vegetation and caused this dune reactivation, here at 75 Mile Beach. Other causes of dune reactivation are believed to be fires and overgrazing. *Photo by Andy Short.*

dune sand. North and South Stradbroke have been particularly heavily mined. When heavy minerals are the sought-after resource, they are separated from the sand on the spot, and the remaining sand is simply put back in place and revegetated. Except for the infuriating habit of engineers to put the sand back in perfectly straight, flat, and featureless ridges that look very artificial, the practice is a good one. After several decades and a few good storms, the dunes and beaches no longer reflect the touch of the engineering profession.

In addition to titanium and zirconium mining, sand is also removed to make glass and for concrete aggregate. Sand used for those purposes, of course, is not returned to the island.

Considerable public pressure is being applied to halt all sand mining on the islands. Protests, sit-ins, and camp-ins have been staged, especially in an effort to stop mining on North Stradbroke Island. A coalition of environmental groups called the Stradbroke Island Action Coalition (SIAC) leads the way. The campaign recently garnered national attention when the chief spokesperson for the environmentalists changed sides and became the chief spokesperson for the mining company!

Fraser Island was designated by UNESCO as a World Heritage site because it is the world's largest sand island and because of its beautiful lakes and streams, fern and palm rain forest and the spectacular dunes. Some of the trees in the forest exceed two hundred feet in height. Among other things, the World Heritage site designation brought a halt to the sand mining that had been going on since the turn of the century.

Like many barrier islands in the world, these islands have had problems with an overabundance of animals whose natural predators are gone. For a while, on Fraser Island, both dingos and feral horses were out of control numerically. The dogs were killing too many wallabies and the horses were overgrazing. Removing animals has solved the problem.

THE FAKES

The Georgia barrier islands are accidents. They exist only because of the strangest of coincidences, the close correspondence in elevation of two successive Ice Age sea levels. Because the two sea levels were close, the barrier islands are close. If today's sea level were a mere ten feet higher or ten feet lower, the composite islands wouldn't exist. There would be only a single island chain.

The future of the Georgia barrier islands, in a time of rising sea level, is continued erosion. The modern or Holocene barrier islands will soon disappear, most within a century or two. Island narrowing of the Pleistocene islands will continue until the point at which the islands are occasionally overwashed across their entire widths by storm waves. At this juncture, the now much narrowed and lowered Pleistocene islands will once again become barrier islands in their own right, and the Georgia coast will be lined by a single chain of narrow islands. The people on the Georgia barrier islands, however, will not take kindly to this natural evolution—larger seawalls, intended to hold the islands in place, are just around the corner.

The Australian islands are also accidents of the level of the sea, but in a different sense than the Georgia islands are. They are the final remnants of a once extensive coastal dune field, which somehow ended up as elongated bodies of sand superficially similar to true barrier islands. There are no other such chains of barriers in the world.

The future for the Queensland islands is their eventual disappearance, perhaps in a few centuries, as gradual erosion continues in a unceasingly rising sea level. Natural evolution in future decades and centuries will probably not be substantially altered by humans, since the islands are not already developed or slated for development. These islands can only get smaller, and they cannot migrate until they are considerably smaller. At some point, when the islands have lost most of their volume and area, and they are low enough for overwash to occur, they, too, will become real and active barrier islands. Just like the Georgia islands, the Aussie islands will once again become one with the sea and begin true island migration.

—

Requiem for Some Friends

IT is not at all a giant leap to liken barrier islands to well-oiled machines. Dozens of parts work together, each dependent upon the other, to achieve the desired motion. An even better analogy for barrier islands is life itself. Living things are smoothly operating systems that evolve in predictable and sensible ways to enhance their survival and their health. And so do barrier islands. Barrier islands lead a Gaia-like existence. The Gaia Hypothesis, briefly stated, holds that all life on Earth is a form of life itself. Earth's entire ecosystem is in a giant global dynamic equilibrium with all the components that are able to adjust to the endlessly changing biological, geological, oceanographic, and atmospheric factors that affect living things.

Barrier islands exist in a similar dynamic equilibrium. Sea level rise moves the shoreline back. Storms raise the elevation and widen the island by overwash. Winds blow sand ashore to form dunes. Dunes and overwash are held in place by vegetation. Salt marsh builds up the lagoon side and widens the island. Inlets come and go, storing and releasing sand. Longshore currents lengthen islands.

Even if the life analogy is reaching a bit, putting barrier islands in a Gaia context does provide a basis for thinking about how to live with them in a way that will keep islands existing and evolving, a way to preserve them for future generations. Perhaps more important, the distinction between life and death, as it applies to a barrier island, should provide guidance to distinguish good and bad development practices.

Once the case is made that barrier islands are analogous to living things, it is then possible to relate individual aspects of island behavior to life processes.

Barrier islands require sustenance. Sand is food to barrier islands.

They use energy. Waves and, to a lesser extent, wind and tides furnish energy to the islands.

They grow fat. A large sand supply and/or stable sea level results in island widening.

They grow thin. Sand starvation or rapid sea level rise thins them down.

They protect themselves. Beaches flatten to absorb storm waves. Islands migrate to avoid inundation by rising sea level.

They recover. Sand comes back to the beaches and dunes after the big storms, islands build up after subsiding, and inlets narrow after blowing out during storms.

They have friends and enemies. Storms are their friends, necessary for their evolution. Their enemies are engineers who arrest their evolution.

They have different personalities. Different combinations of sand supply, sea level change, subsidence, vegetation type, island orientation, and oceanographic settings produce huge variations in modes of island development.

They die. The death of a barrier island is a natural event when sea level drops, when an island migrates up against the mainland, or when sea level rises so fast that the island's sand supply can't keep up. People kill barrier islands by cutting off their sand supply (e.g., jetties at inlets), isolating them from their energy sources (e.g., seawalls), and removing their ability to protect themselves from sea level rise (e.g., removal of storm overwash sand from the island after storms).

HOW TO KILL BARRIER ISLANDS

Historically, the use of barrier islands in the United States, where most of the world's developed islands reside, began with wary, cautious, and knowledgeable use of the islands. Today's approach is no-holds-barred development with little or no recognition of islands' natural histories or the processes at work that keep them alive (fig. 12.1, batik 12.1). It is doubtful if the differences between early cautious development and the later brute force approach are the result of the different sensitivities of the people involved. Most likely the differences are economic, a reflection of the disparity between rich and poor societies.

Among the least intrusive societies on any of the world's developed barrier islands is that on Bazaruto Island, Mozambique, a culture low on the economic scale. The most intrusive is Taiwan, where barrier islands are sacrificed on the altar of industrialization.

On Bazaruto, the population density is low—two thousand people on an eighteen-mile-long island are scattered in small villages, living exclusively in grass huts (fig. 12.2). Sparsity of people, however, isn't the major reason for the very gentle manner in which the barrier island is treated. Bazaruto is inhabited by a subsistence society that depends mostly on the sea for food. Not a single structure, grass hut or otherwise, is located within a half mile of the beach. Grazing by goats is kept at a sustainable level, as is harvesting of the sparse forests. No roads are paved, and a bulldozer's

FIGURE 12.1 *(above)*
The twenty-five-plus-story Point of Americas Condominium adjacent to Port Everglades Inlet, Fort Lauderdale, Florida. Such buildings remove any flexibility of response to the sea level rise and greatly increase the problem of evacuation before major storms strike. *Photo by Charles Finkl.*

FIGURE 12.2
A typical home on Bazaruto Island, with fish-drying racks. The subsistence society of this island moves few dunes and produces little garbage.

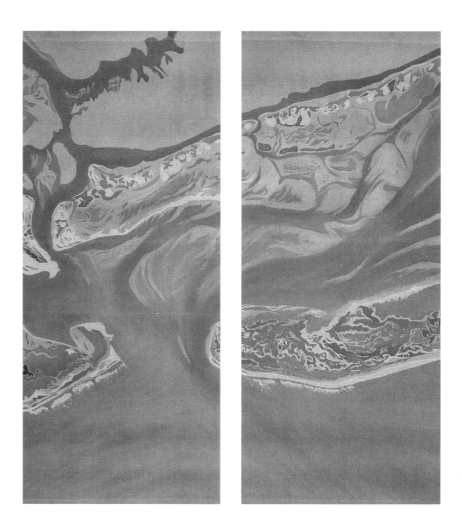

BATIK 12.1
Shackleford Banks, N.C., 1993, 49" x 43"
Beaufort Inlet, which separates
Shackleford Banks and Bogue Banks, has
been dredged for two hundred years. The
sand bodies in the sound behind the inlet
are all part of the flood tidal delta, but
these have been modified by dredging. In
the upper part of the batik is a line of
dredge spoil islands marking the channel
in front of the town of Beaufort.

Artist's note: Beaufort Inlet, a major
shipping channel used by oceangoing
vessels, is visible in this batik. In the
center is the Rachel Carson Estuarine
Preserve, an island consisting of dredge
piles surrounded by channels within the
sound. I enjoyed depicting sand
underwater.

roar has never been heard on the island, except during the construction of
a small tourist airstrip. Shipwreck timbers and other useful items that drift
to the beach are routinely cleaned up. The resorts, two of them, are located on the lagoon shoreline, and their footprints on the sand are light.

No change in this style of living seems likely in the foreseeable future of
Bazaruto. With no wealthy government waiting in the wings to lend a
hand, the cashless economy of Bazaruto is unlikely to impede barrier island
processes. Population is not likely to grow. The local fisheries are already
overextended. Access to the island is possible only by small, overcrowded
dhows, sailing boats of Arab design, leaving island residents to deal with
isolation from friends and family on the mainland.

Isolation, especially in times past, was a hallmark of barrier island living.
Among other things, an all-surrounding body of water leads to the preser-

vation of languages, dialects, and accents. Barrier islands are largely responsible for preserving the Frisian language in Holland, the Gullah dialect in Georgia, and the Outer Banks Elizabethan English accent in North Carolina. Now, with the coming of development, island accents are an endangered species.

The civilizations on barrier islands change as much as the islands themselves. Back when the Inupiat Eskimos, along the Chukchi Sea, were a mobile people, they were able to establish a new settlement almost on a whim. If shorelines retreated or big storms created problems, they could move away to another spot on the island or to the mainland. They were a dynamic people on dynamic islands and were as gentle to their islands as are the natives of Bazaruto. But now the Inupiats are wealthy, or at least they are part of a wealthy society. Now that a modern government requires (and the Eskimos desire as well) waste disposal, heating, good drinking water, and minimum housing standards, the Inupiat are no longer mobile and they no longer live at peace with their barrier islands. Buildings are constructed on lots carved from flattened dunes. Because the buildings are stationary and "permanent," there is an Inupiat erosion problem for the first time (fig. 12.3). The shorelines around the Chukchi Sea villages of Kivilina and Shishmaref are lined with seawalls, groins, and even discarded snowmobiles. Moving villages off the island to the mainland is almost an economic impossibility (cost: $50 million for a village of six hundred individuals).

Buildings in Shishmaref are mostly small, single-family, government-issue homes that, without exception, are easily movable. It is highly unlikely that the Chukchi Sea islands will ever see high-rise condominiums, miniature

FIGURE 12.3
An Inupiat erosion problem. This retreating shoreline is on the lagoon side of the Arctic village of Kivilina, a community that is taking steps to move off the island.

golf courses, or McDonald's restaurants. At best, only a dozen small buildings are adjacent to the beach on Shishmaref, yet these buildings are the only reason there is an erosion problem. Just as surely as massive seawalls on intensely developed New Jersey beaches will destroy islands, the structures in front of this tiny, remote Eskimo village will also kill islands. The U.S. Army Corps of Engineers decides whether to move buildings back or build seawalls, and, absent public concern about beach quality, the corps will always choose seawalls, the most heavy-handed engineering approach. Thus the *strategy* chosen to confront the shoreline retreat problem, rather than development intensity, is what determines the fate of these Arctic barrier islands.

Erosion strategy, in fact, is the principal reason that barrier islands are in trouble around the world. Engineers have responded to the needs of beachfront-property owners rather than to greater societal needs. They have not been constrained by concern for the long-term future of beaches and islands. Historically, engineers have viewed their role as saviors of the threatened structures and defenders of the status quo. If problems arise, future generations can solve them with more engineering. This myopic view is bad enough, but the coastal engineering profession long resisted the notion that seawalls destroy beaches. Some coastal engineering practitioners still deny that seawalls destroy beaches and question the concept of barrier island migration.

In Nigeria, subsistence barrier island societies coexist most unhappily with major industrial development. The problem there is that the Nigerian society does not share its wealth with island dwellers, as U.S. society does with the Inupiat. So the impact of development is entirely negative, on both the island and the people. It's the oil producers that armor the beaches, dredge channels, build jetties, construct refineries and docks, and create pollution. The subsistence societies on the islands are, needless to say, seriously affected by this activity.

In the United States the typical sequence of events on developed barrier islands started with usages as gentle as that of Bazaruto Island, Mozambique, and then proceeded onward and upward. Along the U.S. east coast, New Jersey started extensive island development in the eighteenth century, North Carolina followed in the nineteenth, and Florida began crowding its barrier islands in the twentieth century (batik 12.2).

The New Jersey evolution, from living to dead islands (fig. 12.4), involved gradual occupation of the islands by buildings, seawall placement on all flanks, jetty construction at inlets, and destruction of much of the salt marsh.

BATIK 12.2
Breach Inlet, S.C., 1997, 59″ x 35″
This inlet separates Sullivan's Island (with the irregular shoreline) and Isle of Palms. Sullivan's Island has been accreting for decades as sand piles up behind submerged jetties (for Charleston Harbor) at the south end of the island.

Artist's note: A hazardous inlet for swimming and small craft, Breach Inlet rests between Sullivan's Island and Isle of Palms, northeast of Charleston, South Carolina. The low-altitude batik on silk enhances the turbidity of the water and is an attempt to show the effects of humans.

In Nags Head, North Carolina, marsh destruction is just beginning and no seawalls currently dot the shoreline, but the inexorable march toward the demise of island processes moves rapidly onward in a number of recognizable stages:

Native Americans inhabited the Outer Banks only in the warm season to fish and gather shellfish. Like the Bazaruto Island dwellers of today, they left behind piles of shells known as middens, but little other evidence of their presence.

Early European settlers in the eighteenth and nineteenth centuries sought isolation from the complexities of mainland society—including the law. These pioneers lived cautiously on the lagoon sides of the islands, huddling in maritime forests that buffered the winds of storms.

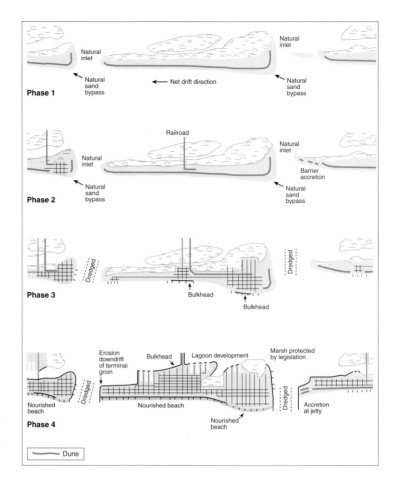

FIGURE 12.4

Sequence of development on a New Jersey barrier island, leading to the complete suppression of the natural processes that once controlled island evolution. *Modified from Nordstrom 2001.*

Tourism began in the mid-nineteenth century. Hotels were then also located on the back side (fig. 12.5), complete with horse-drawn buggies that traveled across the sand on wooden rails to the beach.

Fishing shacks, used for overnight shelter for fishers, were the first "structures" to appear on the beach. Some argue that they could better be called drinking shacks.

The first beachhouses, a row of a dozen or so (fig. 1.5), were built in the late nineteenth century. Many were moved back by mule power from time to time, in one case leaving two underwater brick chimneys in front of the house.

Early beachfront development in the mid-twentieth century often occurred on lots that were six hundred feet deep. The lot size allowed buildings to be moved back as the shoreline retreated.

FIGURE 12.5

Nags Head tourist facilities in the 1920s. From the time of the Civil War on, through the 1930s, buildings used by the tourists were all located on the lagoon side of the island because of the widely recognized storm and shoreline retreat problem. Beach access for tourists was by horse-drawn buggies on wooden rails. Buildings in the early days were not on stilts, and as a result they were often flooded by rising Pamlico Sound storm waters. To prevent their homes from floating off their foundations, residents often had holes or trapdoors in the floors to let the waters pour in. *Photo courtesy of Jack Sandberg.*

The beachfront rush was a post–World War II phenomenon. Beachfront lots became tiny patches of land. Houses were crowded together, and high rises began to appear, providing an even better view of the sea.

A giant beach nourishment project ($1.8 billion for fourteen miles) was approved by Congress in 2000. The cost will amount to $30,000 per beachfront building per year for fifty years.

"WHAT WE NEED IS A GOOD HURRICANE!"

How many times have I heard this expression from people concerned with the direction of barrier island development? It is based on a belief that when the big storm occurs, we'll finally come to our senses and halt the rush to the shore. But it hasn't worked that way. Accelerating property damage in the latter half of the twentieth century (but relatively little loss of life, thanks to an excellent hurricane warning system) seems to have been ignored. Hurricane Hugo (1989) served as an urban renewal project for South Carolina's barrier islands. Small destroyed houses on many islands, such as Folly Beach and Garden City, were replaced by much larger structures. New buildings occupied lots that had lost their pre-storm dunes on shorelines a few feet back from their pre-storm position.

After Hurricane Fran (1997), beachfront lots on Topsail Island, North Carolina, where shallow inlets had opened during the storm, were snapped up by eager buyers (fig. 12.6). Sales were brisk despite plenty of evidence that strong flows had surged across the lots.

It's easy to answer the question of why we have rushed to barrier islands. For poor societies, islands provide a platform close to fishing grounds, afford some protection from marauders and strangers, and furnish a respite from the heat and insects of the interior. For wealthy societies, barrier islands offer the breezes of the sea and proximity to recreational amenities like boating, swimming, and fishing. And rich and economically deprived alike admire their beauty.

But why do we jam buildings right up against island shorelines? The facts that big storms occasionally strike, that shorelines erode, that barrier islands migrate, and that sea levels rise (fig. 12.7) seem not to make the slightest difference to property owners (fig. 12.8). Filling no critical societal needs, such buildings are in fact costly to governments that are required to spend money for their protection. And it is the protection of these buildings that most threatens the existence of barrier islands (appendix, table 4).

Ignorance, irresponsibility, greed, the engineering arrogance of wealthy societies, and political sympathy and support by governments anxious for tax revenue are the usual suspects in the absurd development on the Western world's barrier islands. Bulldozers flatten dunes and forests that might afford storm protection. Canals that provide paths for flooding storm surge waters are cut into islands to give everyone a waterfront lot. "The closer, the better" is the creed, and the better the view of the sea, the more costly the house.

FIGURE 12.6
A shallow inlet on the north end of
Topsail Beach after Hurricane Fran (1997) crossed the island.
A year after the storm, the lots in the inlet were for sale.
Photo by Hugo Valverde.

REQUIEM FOR SOME FRIENDS

Victims of shoreline erosion and storms sometimes exclaim, "How were we supposed to know?" Indeed, the argument could be made that the process of island migration was unrecognized and that erosion rates of most shorelines were unpublished until the 1970s. Perhaps ignorance was fueling the rush to the shore. But why did the Dutch avoid building next to the beach on their islands for hundreds of years? Why did the early settlers on American islands recognize the problem and avoid the beachfront in Nags Head (figs. 1.5 and 12.4)? Why did the first people to build on the exposed beach in Nags Head construct homes that were easily movable? Why were the first lots sold there six hundred feet deep? Why were the buildings in South Seaside, New Jersey, made movable by constructing them on wooden runners? Why do the native people of Nigeria and Colombia still believe in portability (fig. 12.9)? All these people, in all these places and times, apparently knew something that most of today's developed-island dwellers don't?

Arrogance is exemplified by the erection of huge buildings next to rapidly eroding shorelines (fig. 12.1). In 1986, construction workers in Garden City, South Carolina, celebrated the completion of the island's highest high-rise by fishing in the surf zone. The next day they were proudly featured in the local paper as they cast their lines into the sea *from condominium windows*.

Greed is construction of seawalls to protect buildings owned by the few, even though in the long run the seawalls will destroy the beaches used by the many. Sandbridge, Virginia, built a twelve-foot-high vertical steel seawall to save two hundred beach cottages. Originally, both state and local coastal management officialdom turned down the seawall request, so the citizens went directly to the state legislature, which said yes, providing public

FIGURE 12.7 *(left)*
The 1991 Halloween storm (the perfect storm) striking Nags Head, N.C. The entire island at this point is flooded. Most of the buildings shown here were built in the shelter of the old CCC Dune, which has been lost to shoreline retreat. *Photo by Karl Miller.*

FIGURE 12.8
North Sand Key, on the west coast of the Florida Peninsula. At this stage of development, all island processes are halted and the island has become a bar of sand, totally dependent upon humans for its sustenance. *Photo by Richard A. Davis.*

A four-year-old house being moved back from the eroding shoreline on an island in Colombia. The houses in this community are constructed of panels that can be removed and carried to a new location. Poor villages such as these, with no prospect of government aid in any form, adapt to the situation and live flexibly with their barrier islands. *Photo by Bill Neal.*

access was allowed. On my first visit there after the wall was put in, I followed a marked-access path and discovered that a twelve-foot leap was needed to get to the beach. Calling upon my dwindling athletic skills, I made it. Once having arrived, however, I found that I was incapable of scaling the twelve-foot-high smooth wall to return to my car, which was parked illegally because there was no public parking near the public beach access.

Today the Americans who occupy barrier islands are a wealthy and influential lot. Irresponsible resistance to government controls on barrier island development is universal. In the 1980s, South Carolina attempted to prohibit construction of buildings in obviously unsuitable barrier island locations. At the north end of Isle of Palms, the inlet channel wanders back and forth in the well-known fashion of a drumstick island. That's where David Lucas wanted to build a house, at a location frequently flooded by channel meanders. The proposed building site had been underwater three times in the previous forty years, virtually a part of the ebb tidal delta. The much-heralded case went all the way to the U.S. Supreme Court, where the jurists held that if the state wished to prohibit foolish development, it must purchase the property. Inflated beachfront property prices (the Lucas property was worth close to a million dollars) made it impossible to save people from their own foolish actions. The Lucas case was a major blow to responsible barrier island development in South Carolina and perhaps in other states as well.

The State of South Carolina purchased the Lucas property and then turned around and sold it to others. As a result of the intense and protracted legal action, from the bottom to the top of the court system, the dangers of development to this particular beachfront lot must have been the most heavily documented and publicized of any in the world. Despite this, a buyer was easily found. Lucas won, but the scientists were right. Under the category of "justice sometimes prevails," the new owners and their neighbors have spent considerable sums to protect their homes from inlet-caused erosion with trucked-in sand brought in by the Dirt Cheap Trucking Company.

Even in national parks on barrier islands people seem unwilling to concede much to nature. The islands are plagued by communities of wealthy squatters. Another difficult problem for barrier island preservation is the inclusion of towns within park boundaries. Eight communities exist within the boundaries of the Cape Hatteras National Seashore in North Carolina. Naturally the townspeople want to have a say in park management. And their inevitable view that the shorelines should be stabilized so their houses won't fall into the sea clashes with park goals to maintain a natural ecosystem.

Overgrazing in parks, though perhaps less damaging than seawalls, is nonetheless a problem for the health and aesthetics of islands supposedly preserved in their natural states. For example, Fire Island National Seashore in New York has herds of deer far too large for the barrier islands' vegetative cover. The grazing of the deer has diminished the sand-trapping and dune-stabilization function of the vegetation. Animal-rights adherents have prevented the Park Service from transplanting the deer to the mainland, even from sterilizing some of them. To tide them over during the winter, local "animal lovers" put out huge bowls of lettuce and other greens. On Shackleford Banks, North Carolina, the herds of horses have destroyed many acres of salt marsh, which is needed as a nursery environment for marine organisms. These herds were effectively removed from the control of the Park Service and are now controlled by horse lovers, who view the barrier island as a giant pasture (figs. 12.10 and 12.11) rather than a complex ecosystem in which the horses are a single element.

FIGURE 12.10 *(left)*
Overgrazing by herbivores, without predators, is a common problem on barrier islands. Wild horses, here grazing on a sand flat on the lagoon side of Shackleford Banks, N.C., have prevented establishment of a large area of salt marsh. The faint green color on the broad flat is from small shoots of salt-marsh grass, kept from growing by the horses.

FIGURE 12.11
A wild horse in a much denuded maritime forest on Shackleford Banks, N.C. Normally this would be virtually impenetrable forest, but grazing has done away with the small trees. When large trees die or are blown over in hurricanes, they are not replaced because the horses remove young trees. *Photo by Ken Susman.*

FIGURE 12.12 *(above)*
View toward Folly Beach, S.C., from the top of the Morris Island, S.C., Lighthouse. Long abandoned, the lighthouse now stands 1,600 feet out to sea, seaward of Morris Island. It was originally constructed about the same distance behind the ocean shoreline. *Photo by C. W. Evans.*

FIGURE 12.13
Waikiki Beach is one of the most famous beaches in the world, ranking with Miami, Rio de Janeiro, and the French Riviera, all of which are artificial beaches. Most of Waikiki Beach looks like this, having completely disappeared in front of a long line of seawalls, built to protect buildings and parks. The tourist beach, in front of the high-rise buildings in the background, is entirely artificial. It was last replenished by using beach sand from another Hawaiian island. Waikiki Beach is the epitome of the problem of beach and island preservation in a time of rising sea level.

The barrier island process most feared by barrier island dwellers is shoreline erosion (fig. 12.12), the essential ingredient of island migration. Dredging navigation channels through inlets almost always leads to increased erosion. Jetties constructed to make navigation channels cheaper and safer increase erosion rates a lot more.

Worldwide, the most common way to halt retreat is to install some form of shoreline armoring or hardening, usually a seawall (fig. 12.13). Seawalls can be made of anything, ranging from junked cars on Core Banks, North Carolina (fig. 12.14), to smooth, massive concrete edifices on barrier island beaches from Texas to Taiwan to Portugal. At the other end of the spectrum is the single twelve-inch-wide board lying next to a Colombian barrier island beach, put there to reduce overflow of spring tides into the village center. Shoreline stabilization is usually carried out on both sides of developed barrier islands (fig. 12.15).

The damage done by holding an open-ocean shoreline in place is fundamental and irreversible. In fact, many of the long-walled New Jersey is-

lands would simply disappear if suddenly all the seawalls were yanked out and beach nourishment halted. The held-in-place islands are just over-steepened, lifeless piles of sand.

The ultimate in barrier island stabilization may be the planned flood-gates in the three inlets that border Lido and Pellestrina Islands, which protect the Venice, Italy, lagoon. Already engineering has been carried to a stage not seen on any other islands in the world. More than five thousand inhabitants of the town of Malamocca on Lido Island and San Pietro-in-Volta and Pellestrina on Pellestrina Island), are guarded by raised structures on both the lagoon and the ocean sides. Such "diking" is a whole new level of barrier island stabilization. Seawalls and pumped-up beaches provide additional elevation on the ocean side, while raised docks, seawalls, and roads elevate the lagoon shorelines.

Venice, the venerable but sinking old city, on an island within the Lagoon of Venice (fig. 12.16, batik 12.3), floods frequently now, and hotels routinely hand out rubber boots to their tenants, who stroll the city on raised wooden walkways. *Acqua alte*, or high water, occurred 101 times in 1996. It is, however, the rare storm surges that are most feared. Heated public discussions over what to do began after a huge 1966 storm flooded the city and did immense damage. Two very costly alternatives have been proposed. One is to raise the buildings in the city. The second is to construct movable hundred-foot-high gates across the inlets. The gates would normally reside on the floor of the inlets and would be raised at the onset of each storm. If the gates come to be, a new global endpoint in barrier island engineering will have been achieved.

FIGURE 12.14 *(left)*
A seawall made of abandoned cars on Core Banks. The wall, constructed in the 1970s and intended to protect three or four small buildings from overwash, was eventually removed by the National Park Service. The cars were "junkers" used by local fishermen.

FIGURE 12.15
The right way. This salt marsh was built in the mid-1970s to protect a golf course from shoreline erosion on Bogue Bank, N.C. It has grown from a thin band of *Spartina* sprouts to a healthy band of mature salt marsh, bushes, and small trees that is one hundred to two hundred feet wide, and the erosion rate is less than one foot per year. It is likely that the marsh will grow upward as sea level rises.

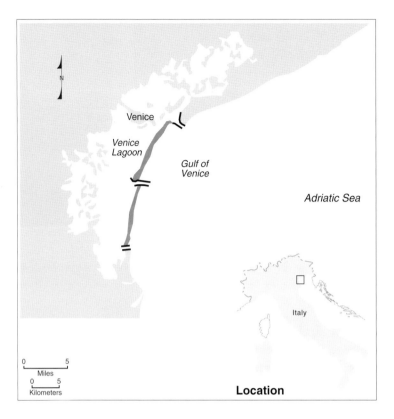

FIGURE 12.16

Index map of the Venice barrier islands and lagoon.

SEA LEVEL RISE AND A CRYSTAL BALL

BATIK 12.3 *(opposite)*

Venice, Italy, 2000, 61″ x 44″

Two barrier islands and two peninsulas enclose Venice and Venice Lagoon. These are perhaps the most armored islands in the world. The entire shoreline is walled, the inlets are jettied, and individual towns on the islands are protected from storm-surge flooding by dikes.

Artist's note: Venice resulted from a journey I made in the summer of 2000 to the famous ancient city. Day trips to the surrounding barrier islands, navigational charts, and satellite images combine to create this batik. On the evenings when we were in Venice, the San Marcos plaza became a shallow lake as the high tides washed over the Grand Canal and flooded through the storm drains.

Assuming that sea level continues to rise as expected, especially if the rate accelerates, in keeping with greenhouse-effect projections, the future of barrier islands around the world will be an intriguing one. Undeveloped island chains, with the notable exception of the Arctic islands, will mostly do just fine (fig. 3.17). The barrier islands treasured by the wealthy Western world, however, are in deep trouble.

No one knows for sure how the plot will play out. For one thing, uncertainties remain over the impact of the greenhouse effect on sea level, ocean warming, storminess, and climate change. For another, much depends on human response to the problem. And that is the least predictable of the variables. It is a certainty, however, that given the opportunity, beachfront owners will protect their property, even if the cost is loss of the beach and cessation of island processes.

The Colombian Pacific coast islands, the Mekong Delta islands, and the Gurupi islands of Brazil should flourish with their huge sand supplies. So,

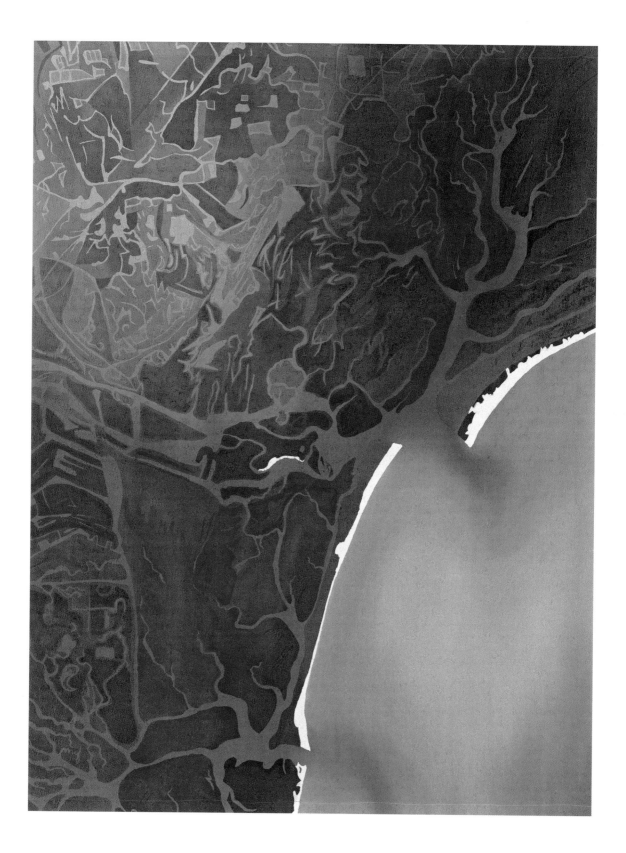

too, should the islands of Mozambique and Abu Dhabi, since beachrock will hold them in place for a long time. Global warming will make the sandur islands of Iceland healthier than ever as glaciers melt and furnish ever more sand to the islands.

I believe the following is a likely scenario for the future of most other barrier islands. Sea level rise coupled with more dam building, channel dredging, and jetty construction will increase the rates of erosion on both sides of barrier islands. That's already happening on undeveloped islands. In response, the islands will become increasingly armored. In North America, beach nourishment will be the most common approach. But in a generation or two this will prove far too costly, especially when artificial beaches begin to disappear at ever-increasing rates as the sea level creeps upward. Armoring will begin in earnest, reaching Taiwan-esque or Venetian proportions.

Even wealthy societies will rebel at the costs of beach nourishment and the widespread loss of the public's beaches because of seawalls. Costs will become prohibitive, and tourist beach communities will be abandoned or left to fend for themselves. Defense of the cities of the lower coastal plain will be the paramount national priority as sea level rises. Manhattan, Boston, Philadelphia, and Miami will become the major concerns, and recreational island property will drop below the horizon of societal importance. The barrier islands will become like Timbalier Island, Louisiana (fig. 12.17). Abandoned seawalls will be isolated in and beyond the surf zone, and de-

FIGURE 12.17
A view of Timbalier Island, Louisiana. There are no residences on this island and no tourist industry. Timbalier Island is home to oil-industry infrastructure and also serves as a wave shelter for a number of drilling rigs behind the island. It has been seawalled at least three times. The first seawall is now almost submerged and is well seaward of the current shoreline. This may be a view of the future of many barrier islands in a time of rapid sea level change. *Photo by Greg Stone.*

struction of buildings will be commonplace with every storm. Meanwhile, the now inflexible barrier islands will have lost their source of energy, their source of sand, and their ability to adapt. They will be lifeless piles of sand.

The situation on our developed barrier islands is amazingly similar to the problems that arose on the Mississippi River. A century of levee building encouraged more development, which resulted in more people and more property in danger from major floods than ever before. The narrowed floodplain of the river actually increased the frequency of floods, leading to four one-hundred-year floods in the last eight years. Beach nourishment on barrier islands also encourages development and will inevitably and quickly lead to more people and property endangered by the prospect of big storms than ever before. And it will kill the barrier islands.

How long will this take? It will, of course, happen at different rates in different places. But within one to four generations—only a few minutes in the life of a barrier island—these dismal scenarios could evolve on islands that are currently developed.

There is a better way. And we don't have to revert to subsistence living as in Mozambique to do it. We can migrate with the islands. This was illustrated by the move of the Cape Hatteras Lighthouse during the summer of 1999 (fig. 12.18). Moving back or off, demolition or surrender to the sea, are virtually the only ways to save barrier islands crowded with buildings. If we choose to save the buildings, all is lost in the long run.

Learning about the processes active on a particular island can provide the basis for development that won't kill the island. A small example: rebuilding roads on top of newly introduced overwash sand rather than removing the sand (figs. 12.19 and 12.20) would allow the island to gain elevation, a critical component of island migration.

FIGURE 12.18
The Cape Hatteras Lighthouse, North Carolina, in its original location. Clearly the lighthouse was in serious danger from the next large storm. The lighthouse was moved back two thousand feet from this location in 1999, an event that had strong repercussions for coastal management thinking in the United States. If a very heavy lighthouse can be moved, why not other buildings threatened by erosion? *Photo by C. W. Evans.*

CANARIES IN THE MINE

Of all the world's major landforms, the barrier islands are perhaps the most sensitive indicators of sea level rise and global warming. Changes that are apparently leading to destruction of the Arctic barrier islands, at least on Alaska's North Slope, may be a direct result of Arctic Ocean warming. If sea level change is playing a role in the Arctic, it seems to be a secondary one to that of warming. The rapid changes in the Arctic islands are occurring in spite of the fact that they are "open" and ice-free only two months of the year.

FIGURE 12.19 (left)
Cleaning the sand off Highway 12 on the North Carolina Outer Banks in the middle of a nor'easter. Although highway workers received justifiable praise for their daring and somewhat dangerous efforts, removing sand is working against the island's efforts to raise its elevation as part of the migration process. Better to let the elevation of the road rise with each storm.

FIGURE 12.20
Rodanthe, N.C., after a storm. Overwash clearly has raised the elevation of the island, an essential element of island evolution in a rising sea level. This sand was eventually pushed back to the beach, returning the elevation of the island to its pre-storm level. A better approach would be to work with the island—raise the street sign and put in a new road using runway matting or gravel. Then do the same after the next storm.

Further south, coastal plain islands are acting as though the sea level is rising. Many, if not most, undeveloped islands are narrowing by erosion on both sides, most likely in preparation for continuing sea level rise. Only narrow islands can migrate rapidly and efficiently, and most coastal barrier islands are too wide to do this at present. Islands today that are actively migrating—Masonboro Island in North Carolina, Assateaque Island in Maryland, Trichen Island in Denmark, and Soldado Island in Colombia—are a few tens of yards wide. That is narrow enough for overwash to occur across an island frequently and on a broad front. On these undeveloped islands, we are witnessing a predictable and beautiful response of nature to sea level rise. They are telling us that Earth is changing in important ways. They are warning us that trouble, deep trouble, lies ahead for civilizations that are wealthy and imprudent enough to build thousands of buildings on these restless ribbons of sand. Nature is patient, nature is powerful, and nature will eventually win at the shoreline. Or as barrier island botanist Paul Godfrey puts it, "Nature always bats last at the shoreline."

Our plea is that barrier islands be recognized as a precious and irreplaceable natural resource. That's the easy part. The hard part is to sense and accept the sacrifices that must be made if we are to save them for future generations. Whether one wants to assume that the islands are well-oiled machines, living things, or just dynamic sandbars, we must exist with them, not on them. We must do so in a way that does not kill them or halt their evolution. The barrier islands of the world are telling us that they need to be free to survive.

accretion: The gradual buildout or seaward growth of a beach or barrier island. Opposite of erosion.

aeolianite: Windblown sand that has been cemented into a rock. Cement is calcium carbonate in the form of the mineral calcite.

barrage lake: The Australian term for a pond on a barrier island that is perched above the water table, held there by a layer of organic matter and clay. Also called "perched lake."

barrier island: A long, narrow sandy or gravel island, bounded by inlets, backed by a lagoon, and facing the open ocean. The islands are typically fifteen to thirty feet thick, a few yards to several miles wide, and may range in length from less than a mile to more than a hundred miles. The seaward boundary of a barrier island is marked by a beach and a shoreface.

barrier island chain: Arbitrarily defined as three or more barrier islands separated from each other and the mainland by inlets but related to one another in terms of sand supply. That is, sand is exchanged from island to island across the inlets.

barrier island migration: The landward movement of an entire barrier island caused by rising sea level and/or a low sand supply.

beach nourishment (replenishment): A means of stabilizing (holding in place) a shoreline by placing sand from an outside source on a beach.

beach ridge: A long ridge of dunes, parallel to the shoreline, that marks the location of the present or a former shoreline. Regressive islands (growing seaward) typically have a series of beach ridges on the surface.

beachrock: Cemented beach sand usually found as a thin (one-to-three-foot) layer between the high and low tide lines of beaches in tropical and subtropical climates. Beachrock layers dip gently toward the ocean and extend along the beach for tens to hundreds of yards. Cement is calcium carbonate in the form of the mineral aragonite.

blowout: An erosional, basin-shaped feature formed by wind. Sand is removed until the water table is reached. Some blowouts become freshwater ponds.

calcareous: Made of calcium carbonate.

caldera: A large and deep, more or less circular, volcanic depression.

chenier: A wooded sand ridge, more or less parallel to the present shoreline, which is a former barrier island, now no longer surrounded by open water.

continental shelf: The very gently sloping surface that extends from the beach to the point offshore (usually between three hundred and six hundred feet water depth) where the slope suddenly increases (the continental slope).

delta: An accumulation of river-derived sediment in a still body of water.

distributary: Stream tributaries flowing away from a river and not returning. Commonly, rivers mouths on deltas break up into a number of distributaries.

downdrift: The dominant direction of sand movement along a beach. Analogous to downstream in rivers.

dredge spoil: Material dredged from navigation channels and harbors.

drumstick island: A common variety of barrier island, shaped like a chicken drumstick with the fat end at the updrift end of the island.

dune: A windblown accumulation of sand.

ebb tidal delta: The body of sand, seaward of an inlet, pushed out by the ebb, or outgoing, tide currents.

erosion: As applied to the shoreline, refers to retreat of the shoreline in a landward direction. Same as shoreline retreat.

estuary: Usually a river valley flooded by the sea, where fresh and salt water mix.

fast ice: Ice attached to the shore that forms each winter and melts during the brief Arctic summer.

fetch: Distance of open sea over which wind can blow to generate waves. The greater the fetch, the greater the potential for high waves.

flood tidal delta: The body of sand, just landward of an inlet, extending into the lagoon, deposited by incoming or flood tidal currents.

frazil ice: Slush in a column of turbulent water in the surf zone. Usually formation of frazil ice marks the beginning of the Arctic winter freezeup.

groin: A wall built perpendicular to the shoreline, usually much smaller than a jetty, that is intended to trap sand to widen a beach and protect adjacent property. The trapped sand usually causes a sediment deficit downdrift and ultimately leads to shoreline erosion.

Holocene epoch: A unit of geologic time extending from 8,000 years ago to the present.

Holocene sea level rise: The rise in sea level after the last major glaciation. The rise began 18,000 to 15,000 years ago.

ice pile-up or ice push: Ice sheet pushed ashore by winds and currents. It buckles and breaks up, forming mounds of sediment-laden ice. When the ice melts, mounds of sediment remain.

ice ride-up: Ice sheet that moves ashore, more or less intact, riding over the barrier island contours.

inlet: A relatively narrow waterway between two barrier islands, connecting the lagoon with the open sea.

jetty: A long wall, on either side of a navigation channel, perpendicular to the shoreline. These are common engineering structures adjacent to barrier island inlets.

jokulhlaup: Icelandic term for a sudden, catastrophic burst of sediment-laden water from a glacier. Same as glacial outburst.

lagoon: A general term for a body of water between barrier islands and the mainland that is relatively shallow and contains salt water. Sometimes referred to as bay, sound, estuary, slough, or bayou.

longshore current: The current, mostly within the surf zone, that transports sand along a beach. The current is caused by waves that strike the shoreline at an angle.

mangrove swamp: Tropical to subtropical marine swamp characterized by mangroves. Rough equivalent of the salt marsh of colder climates. Same as mangrove forest.

NewJerseyization: Destruction of the recreational beach by the long-term halting of shoreline retreat using hard stabilization, i.e., seawalls and groins. Named for New Jersey, where the barrier island beaches have been stabilized for more than 150 years.

overwash: The process of storm waves rolling into the interior of an island, bringing beach sand with them. A very important process in the evolution and migration of barrier islands.

overwash fan: The body of sand, usually long and narrow, deposited on the barrier island by overwash.

pack ice: Sea ice other than fast ice. Pack ice is of many origins and ages, and much of it remains floating on the sea surface year-round.

permafrost: Ground that is permanently frozen a short distance below the surface, usually a few feet at most, on Arctic barrier islands. At a shoreline, the effect of permafrost is to retard erosion.

Pleistocene epoch: A unit of geologic time extending from two to three million years ago to the beginning of the Holocene epoch 8,000 years ago. This is also considered the Ice Age, a time when glaciers retreated and advanced a number of times.

regressive island: An island that within recent history has widened or built out in a seaward direction. A drill hole through the island would penetrate former shoreface sediment.

salt marsh: Mud or sandy mudflat, in protected waters (lagoon) regularly flooded by salt water, covered by salt-tolerant grasses. In temperate North America, a single grass species (*Spartina alterniflora*) makes up the true salt marsh

sandur: Icelandic term for a plain in front of a glacier formed by sediment carried by streams emanating from the glacier.. Also called "outwash plain" or "glacial alluvial fan."

scarp: Small cliff in beach sand that parallels the shoreline. Beach scarps are caused by rapid erosion, often from a recent storm.

sea island: Islands that combine Pleistocene and Holocene islands (Georgia).

seawall: An engineering structure built along a beach, parallel to the shoreline and intended to protect buildings.

shoreface: The narrow, relatively steeply dipping surface extending seaward from the beach, often to a depth of thirty to sixty feet, at which point the slope flattens and merges into the continental shelf. Also called "inner continental shelf." This is the surface of active sand exchange to and from the beach. The depth of the base of the shoreface is controlled by the size of the waves—the larger the waves, the deeper the base.

spit: A long, above-water sand or gravel bar connected to the mainland at one end and terminating at the other end in open water. Spits may become barrier islands when cut through by storms.

spring tide: Tide with higher than normal range, occurring twice a month, corresponding to the full moon.

stamukah: Ice ridges in pack ice formed as ice floes pile into one another.

storm surge: The temporary rise in sea level during a storm, caused mostly by wind pushing water ashore. The raised sea level promotes barrier island overwash.

strudel: A crack in the ice usually formed at the time of spring breakup.

strudel scour: An area of scour on the seafloor caused when fresh water from a river, flowing across the ice at spring thaw time, plunges through a strudel.

subsidence: Sinking of the land surface. A major global cause of sea level rise.

thermokarst lake: A lake or pond formed by summer melting of permafrost and settling of the ground. Same as "thaw lake."

tidal prism: The volume of water that flows in and out of an inlet with the tides.

tide range: The vertical distance between normal high and low tide levels. Same as "tidal amplitude."

transgressive island: A barrier island that has been migrating in a landward direction. A drill hole through such an island would most likely penetrate lagoon sands and muds that the island has migrated across.

tsunami: A sea wave produced by sudden movement of the seafloor, usually associated with an earthquake but sometimes caused by a massive submarine landslide. In shallow water, tsunami waves can reach more than fifty feet in height and can be very damaging to coastal communities.

tundra: An undulating Arctic plain with no trees and a characteristic soil and vegetation assemblage of mosses, lichen, and low shrubs.

updrift: The direction along a beach that is opposite the dominant direction of beach sand movement or littoral drift. Analogous to "upstream" in rivers.

vibracore: A type of sampling device that forces a pipe into the ground using vibration. The sample provides a view of the geologic history of an island.

wave energy: Proportional to wave height. Usually used in reference to the amount of energy expended by breaking waves in the surf zone.

wave height: The vertical distance between the crest and the trough of a wave.

wave refraction: A process of bending of wave crests that occurs as waves come ashore and begin to "feel" the bottom. Wave refraction determines the orientation of breaking wave crests relative to the shoreline.

window lake: An Australian term for barrier island ponds that are at the same level as the groundwater table.

What follows is a list of useful references, both technical and nontechnical, about barrier islands. The list is not a comprehensive one but should provide the interested reader with a foothold in the literature about these islands all around the world. Included are a number of classic papers, those that pulled the science of islands ahead a step or two but that are now a bit out of date.

General

Bird, E. C. F. 1985. *Coastline Changes: A Global Review.* Chichester, U.K.: John Wiley. 219 pp.

———. 1993. *Submerging Coasts: The Effects of a Rising Sea Level on Coastal Environments.* Chichester, U.K.: John Wiley. 184 pp.

Carter, R. W. G. 1988. *Coastal Environments.* London: Academic Press. 615 pp.

Carter, R. W. G., and C. D. Woodroffe, eds. *Coastal Evolution: Late Quaternary Shoreline Morphodynamics.* Cambridge, U.K.: Cambridge University Press. 517 pp.

Coates, D. R. 1972. *Coastal Geomorphology.* Publications in Geomorphology. Binghampton, N.Y.: State University of New York at Binghampton. 404 pp.

Davis, R. A. 1993. *The Evolving Coast.* Scientific American Library. 231 pp.

———. 1994. *Geology of Holocene Barrier Island Systems.* Berlin: Springer-Verlag. 465 pp.

Dean. C. 1999. *Against the Tide: The Battle for America's Beaches.* New York: Columbia University Press.

FitzGerald, D. M., and P. Rosen. 1987. *Glaciated Coasts.* New York: Academic Press. 364 pp.

Fox, W. T. 1983. *At the Sea's Edge.* New York: Prentice-Hall. 317 pp.

Kaufman, W., and O. H. Pilkey. 1983. *The Beaches Are Moving: The Drowning of the American Shoreline.* Durham: Duke University Press. 336 pp.

King, C. A. M. 1972. *Beaches and Coasts.* New York: St. Martin's. 567 pp.

Leatherman, S. P., ed. 1979. *Barrier Islands from the Gulf of St. Lawrence to the Gulf of Mexico.* New York: Academic Press. 325 pp.

———. 1979 (and subsequent editions). *Barrier Island Handbook.* Washington, D.C.: National Park Service. 101 pp.

Nummedal, D. 1985. Barrier islands. In *CRC Handbook of Coastal Processes and Erosion,* 79–121. Boca Raton, Fla.: CRC Press.

Oertel, G. 1985. The barrier island system. *Marine Geology* 63:1–18.

Shepard, F. P., and H. R. Wanless. 1971. *Our Changing Coastlines.* New York: McGraw Hill. 579 pp.

Short, A. 1999. *Handbook of Beach and Shoreface Morphodynamics.* Chichester, U.K.: John Wiley. 379 pp.

Van Rijn, L. C. 1998. *Principles of Coastal Morphology.* Amsterdam: Aqua Publications. 627 pp.

Wells, J. T., and C. Peterson. 1991. *Atlantic and Gulf Coast Barriers: Restless Ribbons of Sand.* Washington, D.C.: National Park Service. 18 pp.

Zenkovitch, V. P. 1967. *Processes of Coastal Development.* J. A. Staers, ed. (English ed.). New York: Wiley Interscience Publishers. 738 pp.

Chapter 1. Dennis Roars Ashore: A Beneficial Catastrophe

Cleary, W. J., and T. P. Marden. 1999. *A Pictorial Atlas of North Carolina Inlets.* North Carolina Sea Grant, Publication UNC-SG 99–04. 47 pp.

Dolan, R., and H. Lions. 1986. *The Outer Banks of North Carolina.* Washington, D.C.: U.S. Government Printing Office. 93 pp.

Godfrey, P. J., and M. M. Godfrey. 1976. *Barrier Island Ecology of Cape Lookout National Seashore and Vicinity, North Carolina.* National Park Service Scientific Monograph Series, no. 9. Washington, D.C.: National Park Service. 160 pp.

Pilkey, O. H., et al. 1998. *The North Carolina Shore and Its Barrier Islands.* Durham: Duke University Press. 318 pp.

Tait, L. S., ed. 1990. *Beaches: Lessons from Hurricane Hugo.* Proceedings of the Third Annual National Beach Preservation Technology Conference, Florida Shore and Beach Preservation Association, Tallahassee. Tallahassee: FSBPA. 391 pp.

Chapter 2. The Global Picture

Cromwell, J. E. 1973. Barrier coast distribution: A world survey. In M. L. Schwartz, ed., *Barrier Islands.* Benchmark Papers in Geology, 9:50. Stroudsburg, Pa.: Douden, Hutchinson, and Ross.

Devoy, R. J. N. 1987. *Sea Surface Studies: A Global View.* London: Croom Helm. 649 pp.

Dolan, R. 1972. Barrier dune system along the Outer Banks of North Carolina: A reappraisal. *Science* 176:280–288.

Ehrenfield, J. G. 1990. Dynamics and processes of barrier island vegetation. *Reviews in Aquatic Sciences* 2. 44 pp.

Glaeser, J. D. 1978. Global distribution of barrier islands in terms of tectonic setting. *Journal of Geology* 86:283–296.

Godfrey, P. J., and M. M. Godfrey. 1976. *Barrier Island Ecology of Cape Lookout National Seashore and Vicinity, North Carolina*. National Park Service Scientific Monograph Series, no. 9. Washington, D.C.: National Park Service. 160 pp.

Hayes, M. O. 1979. Barrier island morphology as a function of tidal and wave regime. In S. P. Leatherman, ed., *Barrier Islands from the Gulf of the St. Lawrence to the Gulf of Mexico*, 1–27. New York: Academic Press.

Henderson, V. W. 1989. Exterior controls on barrier island chain morphology and distribution. Master's thesis, Duke University. 149 pp.

Hoyt, J. 1967. Barrier island formation. *Geological Society of America Bulletin* 78:1125–1136.

Moslow, T. F., and D. J. Colquhoun. 1981. Influence of sea level change on barrier island evolution. *Oceanis* 7:439–454.

Nummedal, D. 1983. Barrier islands. In P. Komar, ed., *CRC Handbook of Coastal Processes and Erosion*, 77–122. Boca Raton, Fla.: CRC Press.

Otvos, E. G. 1977. Development and migration of barrier islands, Northern Gulf of Mexico. *Geological Society of America Bulletin* 81:241–246.

Pierce, J. W. Sediment budget along a barrier island chain. *Sedimentary Geology* 3:5–16.

Pilkey, O. H., et al. 1998. *The North Carolina Shore and Its Barrier Islands*. Durham: Duke University Press. 318 pp.

Schwartz, M. 1971. The multiple causality of barrier islands. *Journal of Geology* 79:91–94.

Swift, D. J. P. 1975. Barrier island genesis: Evidence from the Middle Atlantic shelf of North America. *Sedimentary Geology* 14:1–43.

Van Rijn, L. C. 1998. Morphology of barrier islands and inlets. In *Principles of Coastal Morphology*, 323–362. Amsterdam: Aqua Publications.

Wright, L. D. 1995. *Morphodynamics of Inner Continental Shelves*. Boca Raton, Fla.: CRC Press. 241 pp.

Chapter 3. The American Barrier Island Scene: Hot Dogs and Drumsticks

Bush, D. M., et al. 2001. *Living on the Edge of the Gulf: The West Florida and Alabama Coast*. Durham: Duke University Press. 340 pp.

Brown, J. E. 1991. *Padre Island: The National Seashore*. Tucson: Southwest Parks and Monuments Association. 62 pp.

Chapman, V. J. 1976. *Mangrove Vegetation*. Berlin: J. Cramer. 278 pp.

Clayton, T., et al. *Living with the Georgia Shore*. Durham: Duke University Press. 188 pp.

Davis, R. A. 1994. Barriers of the Florida Gulf Peninsula. In R. A. Davis, ed., *Geology of Holocene Barrier Island Systems*, 167–206. Berlin: Springer-Verlag.

Davis, R. A., and P. J. Barnard. 2000. How anthropogenic factors in the back-barrier area influence tidal inlet stability: Examples from the Gulf Coast of

Florida, USA. In K. Pye and J. R. L. Allen, eds., *Coastal and Estuarine Environments: Sedimentology, Geomorphology, and Geoarcheology*, 293–303. Special Publication 175. London: Geological Society of London.

Doyle, L. J., et al. 1984. *Living with the West Florida Shore*. Durham: Duke University Press. 222 pp.

Elko, N. A. 2001. *Morphodynamics of Caladesi Island, Pinnelas County, Florida*. University of South Florida, Geology Department Field Trip Guide. 17 pp.

Finkl, C. W. 1993. Preemptive strategies for enhanced sand bypassing and beach replenishment activities in Southeast Florida: A geological perspective. *Journal of Coastal Research*. Special issue 18:58–89.

FitzGerald, D. M., and S. Van Heteren. 1999. Classification of paraglacial barrier systems: Coastal New England, USA. *Sedimentology* 46:1063–1108.

Hayes, M. O. 1994. The Georgia Bight Barrier System. In R. A. Davis, ed., *Geology of Holocene Barrier Island Systems*, 233–304. Berlin: Springer-Verlag.

Heron, D., et al. 1984. Holocene sedimentation of a wave-dominated barrier island shoreline: Cape Lookout, North Carolina. *Marine Geology* 60:413–434.

Hoyt, J., and J. Henry. 1971. Origin of capes and shoals along the southeastern coast of the United States. *Bulletin of the Geological Society of America* 82:59–66.

Lennon, G., et al. 1996. *Living with the South Carolina Coast*. Durham: Duke University Press. 241 pp.

Morton, R. A. 1994. Texas Barriers. In R. A. Davis, ed., *Geology of Holocene Barrier Island Systems*, 75–114. Berlin: Springer-Verlag.

Morton, R. A., et al. 1983. *Living with the Texas Shore*. Durham: Duke University Press. 190 pp.

Moslow, T. F., and S. D. Heron. 1994. The Outer Banks of North Carolina. In R. A. Davis, ed., *Geology of Holocene Barrier Island Systems*, 47–74. Berlin: Springer-Verlag.

Nordstrom, K. F., et al. 1986. *Living with the New Jersey Shore*. Durham: Duke University Press. 191 pp.

Oertel, G. F., and J. C. Kraft. 1994. New Jersey and Delmarva Barrier Islands. In R. A. Davis, ed., *Geology of Holocene Barrier Island Systems*, 207–232. Berlin: Springer-Verlag.

Pilkey, O. H., et al. 1984. *Living with the East Florida Shore*. Durham: Duke University Press. 259 pp.

Pilkey, O. H., et al. 1998. *The North Carolina Shore and Its Barrier Islands*. Durham: Duke University Press. 318 pp.

Chapter 4. Barrier Islands and Human Realities: Awash in Politics

Andrade, C. 1992. Tsunami-generated forms in the Algarve barrier islands (South Portugal). *Science of Tsunami Hazards* 106:21–33.

Berendsen, H. J. A. 1998. Birds-eye view of the Rhine-Meuse Delta (the Netherlands). *Journal of Coastal Research* 14:740–752.

Eitner, V. 1996. Geomorphological response of the East Frisian barrier islands to sea level rise: An investigation of past and future evolution. *Geomorphology* 15:57–65.

FitzGerald, D. M., S. Penland, and D. Nummedal. 1984. Control of barrier island shape by inlet sediment bypassing: East Frisian Islands, West Germany. *Marine Geology* 60:355–376.

Jui-Chin C. 2000. Geomorphological change on the Tsengwen coastal plain in Southwestern Taiwan. In O. Slaymaker, ed., *Geomorphology, Human Activity, and Global Environmental Change*, 235–247. New York: John Wiley.

Kelletat, Dieter H. 1995. *Atlas of Coastal Geomorphology and Zonality*. Ft. Lauderdale: Coastal Education and Research Foundation (Special Issue No. 13, 286 pp.).

Lin, T. Y. 1997. Differences of sediments and coastal processes at north and south sides of Ding-Tou–Er Barrier. *Proceedings of the Nineteenth Conference on Ocean Engineering in Republic of China*, 547–552.

Liu, J. T., P. B. Yuan, and J. J. Hung. 1998. The coastal transition at the mouth of a small mountainous stream in Taiwan. *Sedimentology* 45:803–816.

Pilkey, O. H., W. J. Neal, J. H. Monteiro, and J. M. A. Dias. 1989. Algarve Barrier Islands: A non-coastal plain system in Portugal. *Journal of Coastal Research* 5:239–261.

Postma, H. 1996. Sea level rise and the stability of barrier islands. In J. Milliman and B. U. Haq, eds., *Sea Level Rise and Coastal Subsidence*, 269–280. The Hague, Netherlands: Kluwer Academic Publishers.

Chapter 5. Delta Barrier Islands: That Sinking Feeling

Allen, J. R. L. 1965. Coastal geomorphology of Eastern Nigeria beach ridge barrier islands and vegetated tidal flats. *Geologie en Mijnbouw* 44:1–21.

Allersma, E., and W. M. K. Tilmans. 1993. Coastal conditions in West Africa: A review. *Ocean and Coastal Management* 19:199–240.

Allison, M. A. 1998. Geologic framework and environmental status of the Ganges-Brahmaputra Delta. *Journal of Coastal Research* 14:826–836.

Coleman, J. M., H. H. Roberts, and G. W. Stone. 1998. Mississippi River Delta: An overview. *Journal of Coastal Research* 14:698–716.

Fentiman, A. 1996. The anthropology of oil: The impact of the oil industry on a fishing community in the Niger Delta. *Social Justice* 23:87–100.

French, G., L. F. Awosika, and C. E. Ibe. 1995. Sea level rise and Nigeria: Potential impacts and consequences. *Journal of Coastal Research*. Special issue 14:224–242.

Ibe, A. C. 1996. The Niger Delta and sea level rise. In J. Milliman and B. U. Haq, eds., *Sea Level Rise and Coastal Subsidence*, 249–267. The Hague, Netherlands: Kluwer Academic Publishers.

Oyegun, C. U. 1990. The management of coastal zone erosion in Nigeria. *Ocean and Shoreline Management* 14:215–228.

Penland, S., and J. Suter. 1988. Barrier island erosion and protection in Louisiana: A coastal geomorphological perspective. *Transactions of the Gulf Coast Association of Geologic Societies* 38:331–342.

Penland, S., J. R. Suter, and R. Boyd. 1985. Barrier island arcs along abandoned Mississippi River deltas. *Marine Geology* 63:197–233.

Stanley, D. J., and A. G. Warne. 1998. Nile Delta in its destruction phase. *Journal of Coastal Research* 14:794–825.

Walker, H. J. and W. E. Grabau. 1999. World deltas and their evolution. *Acta Geographica Sinica* 54:30–41.

Williams, S. J., et al. 1992. *Atlas of Shoreline Changes in Louisiana from 1853 to 1989*. Miscellaneous Investigations, series 1, 2150 A. Washington, D.C.: U.S. Geological Survey. 64 pp.

Chapter 6. Colombia's Pacific Islands: A Subsiding Tropical Paradise

Martinez, J. O., and J. L. Gonzalez. 1994. *Evolucion Historica de las islas barrer del sector de Buenaventura y el Naya*. *INVEMAR*. Serie Publicaciones Especiales, no. 3. 72 pp.

Martinez, J. O., J. L. Gonzalez, O. H. Pilkey, and W. J. Neal. 1995. Tropical barrier islands of Colombia's Pacific coast. *Journal of Coastal Research* 11:432–453.

———. 2000. Barrier island evolution on the subsiding central Pacific Coast, Colombia, SA. *Journal of Coastal Research* 16:663–674.

West, R. 1957. The Pacific lowlands of Colombia: A Negroid area of American tropics. Ph.D. diss., Louisiana State University, Department of Geography. 340 pp.

Chapter 7. The Carbonate Islands: Tropical Permafrost

Friedman, G. M. 1995. The arid peritidal complex of Abu Dhabi: A historical perspective. *Carbonates and Evaporites* 10:2–7.

Hobday, D. K. 1977. Late Quaternary sedimentary history of Inhaca Island, Mozambique. *Transactions of the Geological Society of South Africa* 80:183–191.

Isphording, W. C. 1975. The Physical Geology of Yucatan: *Transactions of the Gulf Coast Association of Geological Societies* 5.25:231–262.

Purser, B. H., and G. Evans. 1973. Regional sedimentation along the Trucial Coast, Southeast Persian Gulf. In B. H. Purser, ed., *The Persian Gulf*, 212 –239.

Senvano, A., L. Rebelo, and J. Marques. 1997. *Noticia Explicativa da carta geologica da Ilha da Inhaca*. Direccao Nacional de Geologia de Mocambique. 31 pp.

Chapter 8. Lagoon Barriers: The Quiet Ones

Cleary, W. J., P. Hosier, and G. R. Wells. 1979. Genesis and significance of marsh islands within southeastern North Carolina lagoons. *Journal of Sedimentary Petrology* 49:703–710.

Winkler, C. and J. Howard. 1977. Correlation of tectonically deformed shorelines on the southern Atlantic Coastal Plain. *Geology* 5:123–127.

Hine, A. C., and J. Boothroyd. 1978. Morphology, processes, and recent sedimentary history of a glacial outwash plain, Southern Iceland. *Journal of Sedimentary Petrology* 48:901–920.

Nummedal, D., A. C. Hine, and J. Boothroyd. 1987. Holocene evolution of the south-central coast of Iceland. In D. FitzGerald and P. S. Rosen, eds., *Glaciated Coasts*, 115–150. New York: Academic Press.

Oeland, G. 1997. Iceland's trial by fire. *National Geographic* 191:51–71.

Ward, L., M. J. Stephen, and D. Nummedal. 1976. Hydraulics and morphology of glacial outwash distributaries, Skeidararsandur, Iceland. *Journal of Sedimentary Petrology* 46:770–777.

Barnes, P. W., E. Reimnitz, and B. P. Rollyson. 1992. Map showing Beufort Sea coastal erosion and accretion between Flaxman Island and the Canadian border. *U.S. Geological Survey Miscellaneous Report*, Map I-1182-H.

Hequette, A., and M. H. Ruz. 1991. Spit and barrier island migration in the southeastern Canadian Beaufort Sea. *Journal of Coastal Research* 7:677–698.

Hill, P. R., P. W. Barnes, A. Hequette, and M. H. Ruz. 1994. Arctic coastal plain shorelines. In R. W. G. Carter and C. D. Woodroffe, eds., *Coastal Evolution*, 341–372. Cambridge, U.K.: Cambridge University Press.

Hill, P. R., et al. 1990. *Geologic investigations of the Canadian Beaufort Sea Coast*. Geological Survey of Canada, Open File Report 2387. 348 pp.

Reimnitz, E., P. W. Barnes, and J. R. Harper. 1990. A review of beach nourishment from ice transport of shoreface materials, Beaufort Sea. *Journal of Coastal Research* 6:439–470.

Reimnitz, E., and E. Kempema. 1982. Dynamic ice wallow relief of northern Alaska's nearshore. *Journal of Sedimentary Petrology* 52:451–461.

Ruz, M. H., A. Hequette, and P. R. Hill. 1992. A model of coastal evolution in a transgressive thermokarst topography, Canadian Beaufort Sea. *Marine Geology* 106:251–278.

Short, A. D. 1979. Barrier island development along the Alaskan-Yukon coastal plains. *Geological Society of America Bulletin* 90:3–5.

Walker, H. J. 1998. Arctic deltas. *Journal of Coastal Research* 14:718–738.

Zenkovitch, V. P. 1967. *Processes of Coastal Development*. J. A. Staers, ed. (English ed.). New York: Wiley Interscience Publishers. 738 pp.

Bird, E. C. F. 1973. Australian coastal barriers. In M. Schwartz, ed., *Barrier Islands*, 410–426. Stroudsburg, Pa.: Dowden, Hutchinson, and Ross.

Hayes, M. O. 1994. The Georgia Bight Barrier System. In R. A. Davis, ed., *Geology of Holocene Barrier Island Systems*, 233–304. Berlin: Springer-Verlag.

Henry, V. J., C. R. Alexander, and J. Crawford. 1996. Mesotidal barrier, back-barrier, and inlet environments of Sapelo Island, Georgia. *Proceedings International Conference on Tidal Sedimentology.* 36 pp.

Roy, P. S., and R. Boyd. 1996. *Quaternary Geology of Southeast Australia: Field Guide to the Central New South Wales Coast.* International Geological Correlation Program, Project 367. 174 pp.

Schoettle, T. 1996. *A Guide to a Georgia Barrier Island.* St Simons Island, Ga.: Watermarks Publishing. 158 pp.

Sullivan, B. 1988. *Sapelo: A History.* Darien, Ga.: Sapelo Island Research Foundation. 88 pp.

Teal, M., and J. Teal. 1997. *Portrait of an Island.* Athens, Ga.: Brown Thrasher Books. 175 pp.

Thom, B. G. 1984. *Coastal Geomorphology of Australia.* Sydney: Academic Press. 372 pp.

Ward, W. T., and K. G. Grimes. 1987. History of coastal dunes at Triangle Cliff, Fraser Island, Queensland. *Australian Journal of Earth Sciences* 34:325–333.

Young, R. W., and E. A. Bryant. 1992. Catastrophic wave erosion on the southeastern coast of Australia: Impact of the Lanai tsunamis ca. 105 Ka? *Geology* 20:199–202.

Chapter 12. Requiem for Some Friends

Bush, D. M., O. H. Pilkey, and W. J. Neal. 1996. *Living by the Rules of the Sea.* Durham: Duke University Press. 179 pp.

Cencini, C. 1998. Physical processes and human activities in the evolution of the Po Delta, Italy. *Journal of Coastal Research* 14:774–793.

Ministry of Public Works. 1995. *Measures for the Protection of Venice and Its Lagoon.* Water Authority of Venice. 37 pp.

National Research Council. 1987. *Responding to Changes in Sea Level.* Washington, D.C.: National Academy Press. 148 pp.

———. 1995. *Beach Nourishment and Protection.* Washington, D.C.: National Academy Press. 334 pp.

Nordstrom, K. F. 2000. *Beaches and Dunes of Developed Coasts.* Cambridge, U.K.: Cambridge University Press. 338 pp.

Pilkey, O. H., and K. Dixon. 1996. *The Corps and the Shore.* Washington, D.C.: Island Press. 272 pp.

Platt. R. H., et al. 1992. *Coastal Erosion, Has Retreat Sounded?* Institute of Behavioral Science, University of Colorado, Monograph 53. Boulder. 189 pp.

TABLE 1 Global Barrier Island Distribution*

Country	Islands	Length (mi.)	Percent (mi.)
USA	405	3,054	23.43
Mexico	104	1,392	10.68
Russia	226	1,020	7.82
Australia	208	905	6.94
Mozambique	115	563	4.32
Brazil	72	559	4.29
India	45	476	3.65
Madagascar	119	377	2.89
Colombia	63	364	2.79
Nigeria	23	337	2.59
Vietnam	29	310	2.38
Canada	154	294	2.25
China	64	218	1.67
Nicaragua	16	167	1.28
Germany	17	163	1.25
Egypt	11	159	1.22
Iran	54	158	1.21
Mauritania	1	142	1.09
Myanmar	25	128	0.98
Iceland	22	117	0.90
Italy	39	114	0.87
Netherlands	9	102	0.78
Indonesia	36	102	0.78

(Continued)

Country	Islands	Length (mi.)	Percent (mi.)
Sri Lanka	5	99	0.76
Bangladesh	12	99	0.76
Pakistan	28	97	0.74
Gabon	3	87	0.67
Senegal	11	83	0.64
UAE	16	79	0.60
Ukraine	4	72	0.55
Panama	16	71	0.55
Ghana	4	65	0.50
Sierra Leone	5	64	0.49
Spain	8	57	0.44
El Salvador	5	56	0.43
Costa Rica	11	55	0.42
Romania	8	54	0.41
Papua New Guinea	8	53	0.41
Ecuador	8	51	0.39
Denmark	7	45	0.34
Tanzania	13	45	0.34
Greece	14	43	0.33
Guatemala	8	40	0.31
Cote d'Ivoire	2	38	0.29
Guinea Bissau	9	38	0.29
Guinea	6	35	0.27
Angola	4	35	0.27
Portugal	6	35	0.27
France	4	34	0.26
Thailand	6	33	0.25
Cameroon	1	27	0.21
Taiwan	6	24	0.19
Benin	2	24	0.18
Kenya	6	20	0.16
Yemem	4	20	0.15
Tunisia	8	20	0.15
Turkey	9	18	0.13
New Zealand	1	16	0.12
Qatar	4	14	0.11
Libya	2	14	0.11
Djibouti	6	11	0.08
England	3	9	0.07
Gambia	2	8	0.06

(Continued)

Country	Islands	Length (mi.)	Percent (mi.)
Ireland	2	8	0.06
Korea	4	7	0.05
Kuwait	2	6	0.04
Scotland	I	6	0.04
Liberia	I	4	0.03

*Compiled by Matthew L. Stutz

TABLE 2 Distribution of Barrier Island Types[*]

Global

HEMISPHERE	ISLANDS	LENGTH (MI.)	COASTAL PLAIN	DELTA
All	2,152	13,035	72.3%	27.7%
Northern	1,595	10,325	76.4%	23.6%
Southern	557	2,711	56.7%	43.3%

Continent

CONTINENT	ISLANDS	LENGTH (MI.)	COASTAL PLAIN	DELTA
North America	663	4,739	92.4%	7.6%
Asia	530	2,786	64.1%	35.9%
Africa	354	2,196	54.1%	45.9%
South America	199	1,364	45.4%	54.6%
Australia	283	1,076	74.5%	25.5%
Europe	153	874	74.4%	25.6%

Ocean Basin

OCEAN	ISLANDS	LENGTH (MI.)	COASTAL PLAIN	DELTA
Atlantic	506	5,163	76.0%	24.0%
Pacific	448	2,857	61.3%	38.7%
Indian	605	2,676	60.3%	39.7%
Arctic	475	1,635	92.7%	7.3%
Mediterranean	92	436	23.4%	76.6%
Persian Gulf	34	143	69.0%	31.0%
Black Sea	12	126	57.3%	42.7%

Tectonic Setting

MARGIN	ISLANDS	LENGTH (MI.)	COASTAL PLAIN	DELTA
Trailing-Amero	894	4,803	79.7%	20.3%
Trailing-Afro	518	3,643	54.6%	45.4%
Marginal Sea	370	2,302	79.2%	20.8%
Collision	321	1,675	59.8%	40.2%
Trailing-Neo	57	486	6.3%	93.7%
Black Sea	12	126	57.3%	42.7

[*]*Compiled by Matthew L. Stutz*

TABLE 3 Arctic Barrier Island Distribution[*]

Country	Islands	Length (mi.)
Russia	203	897
Alaska	142	535
Canada	130	202

[*]*Compiled by Matthew L. Stutz*

TABLE 4 **Endangered and Threatened Species That Utilize Barrier Island Habitats, Listed by the U.S. Government (courtesy of Sidney Maddock)**

Mammals
Alabama beach mouse (*Peromyscus polionotus ammobates*)
Anastasia Island beach mouse (*Peromyscus polionotus phasma*)
Choctawahatchee beach mouse (*Peromyscus polionotus allophrys*)
Perdido Key beach mouse (*Peromyscus polionotus trissyllepsis*)
St. Andrews beach mouse (*Peromyscus polionotus peninsularis*)
Southeastern beach mouse (*Peromyscus polionotus niveiventris*)

Birds
piping plover *(Charadrius melodus)*
roseate tern (*Sterna dougallii dougallii*)

Reptiles
leatherback sea turtle (*Dermochelys coriacea*)
hawksbill sea turtle (*Eretmochelys imbricata*)
Kemp's ridley sea turtle (*Lepidochelys kempii*)
green sea turtle (*Chelonia mydas*)
Atlantic loggerhead sea turtle (*Caretta caretta*)
Invertebrates
northeastern beach tiger beetle (*Cicindela dorsalis dorsalis*)

Plants
seabeach amaranth (*Amaranthus pumilius*)
beach jacquemontia (*Jacquemontia reclinata*)

regressive islands, 38–39, 87;
    Colombia, 153; Galveston
    Island, 38; Yucatán, 164
Reykjavik, Iceland, 212
Reynold, R. J., 251
Riggs, Stan, 184–85
Rio Lagartos Natural Park,
    Mexico, 163
Rio Patia Delta, Colombia, 141
Rio Raposa, Colombia, 150
Roanoke Island, N.C., 183–86
"rock-cored" islands, 72
Rodanthe, N.C., 23, 63
Rosetta Distributary, Egypt, 128

Sabkah, Abu Dhabi, 171
Saffir-Simpson scale, 13
Saint Catherines Island, Ga., 246–48
Saint Helena Island, S.C., 193
Saint Simons Island, Ga., 245, 250
Salazar, Antonio, 175
salt marsh (*Spartina alterniflora*), 73–74
salt production (Yucatán), 163, 166
sand mining, 126
sand supply, 38–40; Louisiana, 114;
    Nigeria, 126; Nile Delta, 127
Sandbridge, Va., 273
sandur, 206, 209–210, 212–13,
    218–19; Breidarnerkursandur,
    210, 212; Myradalssandur, 218;
    Skeidararsandur 206, 208, 214,
    216–17
Sandy Point, S.C., 244
San Juan de la Costa, Colombia,
    141–42, 144
Sapelo Island, Ga., 243, 246, 251–53
Sargent Beach, Tx., 84
Sarichef Island, Alaska, 237
Sea Islands, 46, 245, 251
sea level rise, 40; Holocene, 42;
    Louisiana, 114; New Zealand,
    40; Northern and Southern
    Hemisphere, 40–42;
    Taiwan, 40

Seabright, N.J., 57–58
seawalls. *See* armored shorelines
Severnya Island, Siberia, 232
Shackleford Banks, N.C., 22–24, 40,
    48, 256, 275
shamals (desert windstorms), 167, 171
ship ballast islands, 203
Ship Shoal, La., 121
Shishmaref, Alaska, 235–36, 238,
    267–68
shoreface, 44–45
shrimp farms, Vietnam, 134
sinkholes (ceynotes), 160–62
Skeidararjokull Glacier,
    Iceland, 219
Skidaway Island, Ga., 193
slave trade (Nigeria), 123
Smith, Quentin, 199
Soldado Island, Colombia, 43,
    153–57
Soldado Village, 155
Solomon, Steven, 223
South Seaside, N.J., 273
Southern Hemisphere, 29, 42
Southern Ocean, 256
Spalding, Thomas, 251–52
spoonbill, black-faced, 110
squatters, 100–101
stamukah (ice ridges), 227
Stanley, Daniel, 128
storm rings, 237
Stradbroke Island, North and
    South, 254, 256, 258
Stradbroke Island Action
    Coalition, 260
strudel (crack), 225–26; strudel
    scours, 225–26
subsidence: Nigeria, 127;
    Louisiana, 118
Suffolk Scarp, N.C., 196

Taiwan barrier islands, 104–111,
    280; history, 105–106; sea
    level history, 42

308   INDEX